WEEDS IN WINTER

Weeds in Winter

WRITTEN AND ILLUSTRATED BY

Lauren Brown

HOUGHTON MIFFLIN COMPANY BOSTON 1977

Library of Congress Cataloging in Publication Data

Brown, Lauren.   Weeds in winter.
Reprint of the ed. published by Norton, New York.
Bibliography: p.   Includes index.
    1. Plants in winter—Northeastern States—Identifica-
tion. 2. Weeds—Northeastern States—Identification.
3. Wild flowers—Northeastern States—Identification.
I. Title.   [QK118.B7 1977]   582'.12'0974   77–4712
ISBN 0–395–25785–9 pbk.

Printed in the United States of America

Q 10 9 8 7 6 5 4 3 2 1

To all my friends

ACKNOWLEDGMENTS

I owe more than I can ever realize to Tom Siccama, a former professor at the Yale School of Forestry and Environmental Studies. As a teacher always willing to give time to his students, Tom helped me identify many of these plants and showed me how to go about identifying them on my own. He was genuinely interested in this project, and his interest encouraged me and, most important, got me moving. Dr. Alfred E. Schuyler, of the Academy of Natural Sciences of Philadelphia, who taught me plant taxonomy in the first place, critically read the entire manuscript and corrected many mistakes. I am grateful to him for setting aside so much time on so little notice. Dr. John Mickel, of the New York Botanical Garden, offered helpful suggestions on the section on ferns.

At W. W. Norton & Company, Ned Arnold originally accepted the manuscript, and I owe thanks to Bob Pyle for steering me in his direction. The book was guided through the production process by Sherry Huber, whose efficiency, competence, and pleasantness were admirable, and critical in getting the book out. The designer was Marjorie Flock, who was always a pleasure to work with.

Aside from professional help, several people fall into the moral support category: Lloyd Irland, and my parents, Ralph and Elizabeth Mills Brown. Lloyd and my mother not only read and criticized but also always encouraged me. Liz Mikols, my friend, also read the manuscript and made many valuable comments. Kristine Ashwood saved the day when I needed a typist, and she served as much more, running errands and dropping everything to work on this project. Gary Wolfe generously gave his time as a guinea pig to test out the key. Many others helped me in this regard: my aunt Sarah McCulloch, Anne Bittker, several patient friends, and many students who had no choice, but who were cheerful about it anyway.

LAUREN BROWN

Guilford, Connecticut

CONTENTS

WEEDS IN WINTER

INTRODUCTION

In New England, almost half the year is winter. The flowers and the greenery that brought us so much pleasure from April to October are gone, and people who enjoy looking at and identifying plants put their books away. However, the plants have by no means disappeared and it is still perfectly possible to identify them. Many books have been published already which teach people how to identify trees and shrubs in winter by the characteristics of their twigs and bark. But what about the rest of the plants—the weeds and wildflowers?

Any plant that is not a tree or a shrub is herbaceous—its living parts die to the ground by the end of each growing season. Herbaceous plants include what we commonly refer to as wildflowers, weeds, grasses, ferns, and mosses. Herbaceous plants have three kinds of life cycles. If the entire plant—above and underground parts—dies at the end of each season, it is called an annual; if the underground parts survive and send up new shoots in the spring, it is a perennial. If the plant has a two-year life cycle, it is a biennial. In its first year a biennial produces only a rosette of basal leaves—almost stalkless leaves close to the ground. The second year, the plant sends up a flowering shoot, and then it dies. The garden carrot, which is the same species as the wildflower Queen Anne's Lace, is a biennial.

Although they may be dead, many herbaceous plants do not disappear over the winter. They remain as dead, woody tissue, various shades of brown or gray, sometimes with fruits, sometimes just as stalks; and many of them are spectacularly beautiful. Thoreau wrote about these dried plants in his journal:

> *When the ground was partially bare of snow, and a few warm days had dried its surface somewhat, it was pleasant to compare the first tender signs of the infant year just peeping forth with the stately beauty of the withered vegetation which had withstood the winter—life-everlasting, goldenrods, pinweeds, and graceful wild grasses, more obvious and interesting frequently than in summer even, as if their beauty was not ripe till*

then, even cotton-grass, cattails, mulleins, johns-wort, hard-hack, meadowsweet, and other strong-stemmed plants, those unexhausted granaries which entertain the earliest birds—decent weeds at least, which widowed nature wears.

Many of them bear little resemblance to the plants that they were in the summer, for some of the most inconspicuous summer plants leave the most beautiful remains, and vice versa. The wild yam (*Dioscorea villosa*), for instance, blends in completely with its surroundings in summer, but in winter anyone with an eye for form cannot help but notice its amazing three-sided hearts twining in among other dead plants. On the other hand, many of our more common wildflowers—violets, dandelions, buttercups, clovers—disappear completely by the end of the summer.

You have probably picked these "dried things" and brought them into your house for decoration. You may even have gone to a gift store and paid outrageous prices for something that you could have picked along the edge of the New Jersey Turnpike. If you ever tried to identify any of these plants, you probably found your books useless. By keeping your eyes open, and waiting six months, you might notice what comes up in the spring in the spot where you saw a certain dried stalk in the fall. But it is hard to keep your eyes open all of the time. This book is designed to identify common dried herbaceous plants as they are found in the winter. It is written for the lay person. We assume that you are interested in plants, but that you know little about botany.

You should know a little bit about how plants are classified. They are classified at many different levels, but we will mainly be concerned with three groupings: the *family,* the *genus,* and the *species.*

Family • A knowledge of plant families is extraordinarily useful. Plants are placed in various families on the basis of the characteristics of their flowers. Of course, you cannot see the flowers in the winter, but there are many other family characteristics, such as leaf arrangement, flower arrangement, or kind of fruit, that carry over into the winter. Most plant books are organized by family, so if you know what family a plant is in, identification is a lot easier.

This book is organized for the most part by family. When there is only one representative of a family, or when two winter plants of the same family have nothing in common, the plants have been put in a "Miscellaneous" section at the end. But it should not take you long to recognize some of the more common families, and then you will have a much easier time identifying your winter plants.

Genus and Species • A genus (the plural is *genera*) is a smaller grouping of plants within a family. Goldenrod, *Solidago,* is a genus. A particular kind of Goldenrod, such as Grass-leaved Goldenrod, is the species—*Solidago graminifolia.* Once we have referred to a certain species of *Solidago* in a passage, we can abbreviate it the next time as *S. graminifolia;* if we want to refer to another species in the same genus, we can call it *S. rugosa,* or *S. nemoralis.* You will see some plants in this book identified by the genus name followed by *spp.*—for example, *Solidago spp.* This designation indicates that we have an unidentified species of the genus *Solidago.*

Do not be afraid of Latin names. They eliminate much confusion, because a specific plant has only one Latin name, which applies to that plant only. Common names have developed informally and they vary widely, so that misunderstandings often arise. For instance, *Lepidium campestre,* a member of the Mustard family, in three widely used wildflower guides is called Field Cress, Cow Cress, and Field Peppergrass. Some of its close relatives are Spring Cress, Winter Cress, Early Winter Cress, Rock Cress, Peppergrass, and Field Pennycress. It becomes hard to remember which Cress is which. *Lepidium campestre,* however, is always *Lepidium campestre.* The Latin names may seem meaningless, but if you know any Latin derivatives, you will see that many of them are very descriptive. I have tried to explain the Latin names wherever they are helpful.

For most of the scientific names, this book uses the nomenclature of *Gray's Manual of Botany,* Eighth Edition, by M. L. Fernald. For common names, I have followed various authorities. Other standard information about the plants, such as height, range, and habitat, has been taken from *Gray's Manual;* from *The New Britton and Brown Illustrated Flora of the Northeastern United States and Adjacent Canada,* by H. A. Gleason (hereafter referred to as "Britton and Brown"); from *Wild Flowers of the United States: The Northeastern States,* by H. W. Rickett; and from *A Field Guide to Wildflowers,* by R. T. Peterson and M. McKenny.

HOW TO IDENTIFY DRIED PLANTS

Dried plants are not as easy to identify as green ones. We often tell live plants apart by the color of their flowers and their season of blooming, but neither of these characteristics is much help for gray-brown stalks that are dead from October to April.

Blooming plants are also often identified by the technical differences in the structure of the flowers. This is fairly straightforward; once you know the names of the flower parts, you can easily compare their size, arrangement, and relative numbers. Since flowers become fruits, it should be easy to apply the same technique to winter plants, but for the lay person it is not.

First of all, the dried plants you might find do not necessarily have any fruits at all. The fruits might fall off early in the season and leave a beautiful structure that you want to identify. Second, it is not obvious what a fruit is. Technically, a fruit is a ripened ovary. The ovary is the part of the plant that contains the ovules, and the ovules contain the female reproductive cells. A fertilized ovule becomes a seed, and the ovary ripens around it to become a fruit. Fruits contain seeds. If you think about an apple, or a watermelon, this makes sense. However, most of the fruits in this book do not look like apples or watermelons, and if they did, they would have rotted or been eaten before we could get out to pick them. Most of the fruits described in this book are small and woody. Some of them, though, are still identifiably fruits, especially if they have seeds inside them. But some are small, hard, dry, and fused to their seeds, so that you cannot tell where the fruit ends and the seed begins. What you and I might call a seed might actually be a fruit, as in the Daisy family (p. 167) and the Parsley family (p. 118), and what we might call a fruit sometimes turns out to be a seed (Wild Leek, p. 59). What we might think is a fruit might also be dried bracts or sepals.

Therefore, it is obvious that conventional botanical terminology is not very helpful for a lay person trying to identify dried winter weeds. However, there are many ways around this problem, and there are still many easy ways to identify dried plants. This book has a key, the use of which is explained on page 18. Step by step, the key points out differences that enable you to identify your plant. You might prefer not to use the key and just to flip through the pages. This is a time-honored method of identifying plants, but your flipping will be more productive if you look for clues first. With a minute's worth of looking, you might notice a characteristic that will tip you off to the identity of your plant.

I. SMELL THE PLANT

Crush the fruit, the dried-up flowers, the stem, the root, or the basal leaves. Does the plant smell like mint? Then it is probably in the Mint family, especially if it has a square stem. Turn to page 137. Does it smell like parsley or carrots? Then it is probably in the Parsley family, especially if the flowers are in an inverted umbrella shape (an umbel). See page 118. Does it smell smoky? Try the Daisy family (p. 167). Does it smell like burning rubber? Try the Tomato family (p. 150).

2. LOOK AT THE LEAVES

Although most leaves die in the winter, they can still be useful in identification. First of all, some biennials and perennials have basal leaves that survive the winter, protected by snow and dead plant matter. If you find the basal leaves, and if you are familiar with plants in summer, you might recognize the plant. The other leaves on the stalk, although dead and shrivelled, often keep their original shape, and if you unfold them and look at them closely, you might recognize them, too. Even if the leaves fall off, they leave a scar at the point where they were attached, and the arrangement of the scars can help you identify the plants. The Glossary shows the most common types of leaves and the different ways they are arranged on the stem. This arrangement can be characteristic of a family. For instance, members of the Buckwheat family and the Parsley family always have wraparound leaves, and all members of the Mint family and the Pink family have opposite leaves.

3. LOOK AT THE STEM

Is it smooth or fuzzy? Shiny or dull? Does it have thorns? What shape is it in cross-section? If it is triangular, you might have found a sedge, which would be beyond the scope of this book. If it is square, there is a good chance that it is in the Mint family.

4. NOTICE WHERE THE PLANT IS GROWING

Is it in a swamp or on a dry ridge-top? In a meadow or in a forest? Is it in a hardwood forest (oak, maple, etc.) or is the forest mainly conifers (pine, hemlock, and the like)? Is it in a vacant lot? On the side of the road? Many plants are very particular about their surroundings, and certain plants turn up predictably in certain places. You cannot identify a plant from its habitat alone, but the habitat can help you to confirm your decisions.

5. DO NOT TAKE JUST ONE PLANT

If there are many of the same plant, do not just grab one and leave. Look around at all of them and make sure you have found one with as many clues as possible.

6. LOOK HARD AT THE PLANT

The more you look, the more you will see.

🌷 🌷 🌷

HOW TO USE THE KEY

At each step of the key, you have to make a choice between two alternatives. You always start with entry #1 and decide whether your plant is a vine (1a) or free-standing (1b). Make your choice and proceed to the paragraph indicated with a boldface number. When you find this entry you will again be faced with two choices. The appropriate choice will lead you to yet another entry, and so on. Through this process you will arrive at an answer with a page number next to it. Turn to this page and compare the drawing and the description with your plant. If they match, you have identified your plant. If they are totally dissimilar, you have probably made a mistake somewhere in the key. Go backwards from your ending point until you reach a point where you might not have been sure, and take the alternate path there.

As a simplified example, let us key out Horse Nettle, pictured here:

> 1a Plant does not stand erect; it is either prostrate or a climbing vine. . . . 2
>
> 1b Plant is not prostrate nor a vine, but stands by itself. . . . 11

The plant is not a vine, so you choose 1b and go to Step 11.

> 11a Plant has no visible fruit, calyx, bracts, seeds, or anything that looks like fruit. . . . 100
>
> 11b Plant has some kind of definable structure, whether it be a true fruit, a cluster of seeds, a lot of fuzzy hairs, a dried calyx, or dried bracts. . . . 12

The plant has a definable remaining structure, so you choose 11b and go to Step 12.

> 12a Fruit is a berry; fleshy. . . . 13
>
> 12b Fruit is dry, not a berry; woody or papery. . . . 15

The fruit is a berry, so you choose 12a and go to Step 13.

> 13a Fruits are yellow; stems might have thorns.   Horse Nettle, p. 151.

If you turn to page 151 you will see a picture that looks like Horse Nettle, and you will read a description that matches it. You have identified it. In some cases the key will lead you to a species name; in other cases it will lead you to a family name. If you come to a family name, turn to the page indicated, read the family description to make sure you are on the right track, and then flip through the pages till you find your plant.

There are several commonly used words in the key of which you will probably think you know the meaning. However, in botanical language they have a specific meaning which is presented in the glossary. If you are not already familiar with botanical terminology, be sure to look up any word printed in CAPITALS. Otherwise, you might operate under a misconception which could lead you down the wrong path. If you have a preconceived notion of what your plant is, check it out first. If your idea is wrong, you will get quite confused using the key. If possible, use the key with a friend. It is much easier if you have someone with whom to discuss your choices.

Read each choice before you make your decision, even if you think you know the answer. If you are not sure what choice to make in the key, do not worry about it. Many of the characteristics of these plants are ambiguous, and the key is designed so that either choice will get you the right answer. Do not brood and nit-pick over your choices; your first instincts will usually be right. Note any steps where you are not sure, and go back to these if you get a wrong answer.

IF YOU DO NOT FIND SOMETHING IN THIS BOOK

This book is an introduction for the non-specialist. It includes the most common and the most spectacular plants—that is, the ones you are the most likely to notice. Most of the plants in this book grow throughout and even beyond the northeastern quarter of the U.S.—from the Atlantic Coast west to Minnesota, Iowa, and Missouri, and from Quebec and Ontario on the north to Virginia, Kentucky, and Missouri on the south. If you live in any of the states within these boundaries, or even slightly beyond, this book will be useful. However, the book is based mainly on observations of the flora of the East Coast. Although almost everything in it grows in the Midwest, you might also find some midwestern species which are not in the book. Also, plant distribution is patchy, and you might just happen to live by an entire field of a plant that is found in very few other places. There are other reasons why you may find a particular dried plant missing from this guide.

One reason might be that you have picked a twig from a tree or shrub. Scratch the stem soon after you pick it. If the inside is green, rather than brown, the plant is still alive, and it is a shrub. This book does not cover shrubs (with a few exceptions, such as *Spiraea* and Bittersweet, which are commonly considered "wildflowers" because of their small size). However, you can identify most shrubs and trees very easily with an inexpensive little book by William Harlow, *Fruit Key and Twig Key to Trees and Shrubs* (Dover).

If you try to identify all of the dried plants in the arrangement on your grandmother's mantle, you might find something that she bought at the florist's shop that does not grow in the northeastern part of the country. The eucalyptus, for example, grows in California, and its round, clasping coppery leaves are often sold for decoration. Moreover, many of the dried "plants" in formal arrangements are not plants at all, but are made from wood shavings, pieces of straw, etc. Garden plants are not covered in this book either, but once you learn how to look at dried plants, you will easily be able to tell which dried stalk in your garden belongs to which flower.

You might also have found a gall. Galls are deformations caused in plant tissue when certain insects lay their eggs in or on the plant. Usually the larva develops in the gall and uses the plant tissue for food or for shelter. Some galls are remarkably complex and beautiful. Surprisingly little is known about galls, but one very thorough book is *Plant Galls and Gall Makers,* by E. P. Felt (Hafner, 1965). If you can find this book, it has very good pictures and should help you to identify the more common galls.

Your plant might be battered beyond recognition by the weather, especially if you find it late in the winter. I have tried not to choose perfect specimens for the drawings since in nature few plants are perfect. However, if your specimen is really mangled, wait until next winter to find a better one, or use your imagination when looking through the book.

If you want to try your own hand at identifying something that is not in this book, we propose various methods, none of them foolproof, but all worth a try. One thing to do is to go back to the site where you found the plant in the spring and summer. If you want to catch the plant in flower, you should go back frequently or you might miss it. If the plant is an annual, it might not be growing there any more, but its seeds might have germinated nearby, so roam around a bit.

If your dried plant still has seeds, you could plant some and see what comes up. Although this might be an educational experience, it is not a good way to identify dried things. First, your chances of germination are small. Wild seeds need various treatments to stimulate off-season germination, and it is hard to know what kind of seed needs what kind of treatment. Sometimes a scratch on the surface will do the trick, or a few days of freezing. Some seeds will germinate if they are soaked for a few days on a wet rag or paper towel, but if they get too wet, they will be attacked by a fungus and rot. Some seeds have to be boiled. And you might simply get a bad seed that would not germinate under any conditions.

Even if your seeds do sprout, you must be sure to plant them in sterilized commercial potting soil; otherwise you will have every weed imaginable growing in your pot. Since indoor conditions in winter are different from outdoor conditions in spring, your plant might come up deformed, and you will be unable to recognize it. If it is a biennial, you will get nothing but leaves the first year, which you might have a hard time identifying. I do not recommend this method.

The best method is to get access to a copy of *Britton and Brown.* For this method to work, you must have a good idea of what family your plant is in. Britton and Brown has an exquisite line drawing of every species in the Northeast, along with a written description. So turn to the family you think your plant is in, and just start looking at the pictures. Your aim is to try to eliminate species until you find one that must be your plant. For each species, ask yourself the following questions:

• What size is the plant? What growth form (vine, crawls along the ground, etc.)?
• What kind of fruit does the plant have? *Britton and Brown* often has pictures of the fruit, especially for families like the Mustard family. If there is no picture, the fruit is described in words.
• What kind of branching does the plant have? Are the flower stalks opposite or alternate? in whorls? in an umbel? Are they short or long? Do they curve up or droop down?
• Where does the plant grow? If you are in Connecticut, and the written description says that a plant grows only in Kansas, then this is probably not your plant.
• Are any dried leaves still on the stem? Unfold them and see what they look like.

If you find a plant that looks like yours, read the written description (with the help of a glossary and metric ruler) to try to confirm your guess. Look for characteristics such as hairiness or stickiness. Once you think you have it identified, you should still try to field check to make sure.

KEY

1a Plant does not stand erect; it is either prostrate or a climbing vine. . . . **2**

1b Plant is not prostrate nor a vine, but stands by itself. . . . **11**

2a Plant has many small hard fruits (ACHENES), each one with a long silky plume.   Virgin's Bower, p. 83

2b Plant is not as above. . . . **3**

3a Fruit stalks are in an UMBEL.   Carrion-Flower, p. 60

3b Fruit stalks are not in an umbel. . . . **4**

4a Fruits have a fleshy red inside with a yellow or orange covering.
                    Bittersweet, p. 214

4b Fruits are not as above—they are dry, woody, or papery. . . . **5**

5a Fruits are tiny (less than ¼″ wide), heart-shaped, and flattened.
                    Common Speedwell, p. 159

5b Plant does not have the above combination of characteristics. . . . **6**

6a Fruits are opened. . . . **7**

6b Fruits are not opened. . . . **10**

7a Fruits are dotted with weak thorns.   Wild Cucumber, p. 240

7b Fruits have no thorns. . . . **8**

8a Fruits are three-sided, each side more or less heart-shaped.
                    Wild Yam, p. 206

8b Fruits are not as above. . . . **9**

9a Fruit is a tan pod twisted into a spiral; stem smooth.

Everlasting Pea, p. 112

9b Fruit is a gray, papery pod; stem shreddy.

Black Swallow-wort, p. 133

10a Fruits consist of round HEADS of small pointed achenes.

Bur-reed, p. 202

10b Fruit is not as above; smooth and rounded.

Morning-Glory family, p. 134

11a Plant has no visible FRUIT, CALYX, BRACTS, SEEDS, or anything that looks like fruit; merely has stubs or STALKS, dried
leaves or vague reminders of fruits or flowers. . . . 100

11b Plant has some kind of definable structure, whether it be a true
fruit, a cluster of seeds, a lot of fuzzy hairs, a dried calyx, or dried
bracts. . . . 12

12a Plant has berries; fleshy fruits. . . . 13

12b Fruits, or other remains, are dry—woody or papery. . . . 15

13a Fruits are yellow; STEM might have thorns.   Horse Nettle, p. 151

13b Fruits are blue or black. . . . 14

14a Fruits are in an umbel.   Lily family, p. 56

14b Fruits are not in an umbel.   Blue Cohosh, p. 208

15a Tiny, fluffy fruits are in a cigar-shaped SPIKE.   Cat-tail, p. 200

15b Plant is not as above. . . . 16

16a Plant is a grass, no matter what size. If leaves are still on the stem,
they are long and pointed, and the stem is straw-like and hollow at
the nodes.   Grass family, p. 50

16b Plant is not a grass. . . . 17

17a Plant has BURS. . . . 18

17b Plant has no burs. . . . 23

|||||||||||||||||||||||||||||||||||||||||||||||||||||||||||||||||||||||||||||||||||

0   1   2   3   4

18a Fruits are flat and borne singly or attached to each other in a chain.
Tick Trefoil, p. 108
18b Fruits are not as above. . . . **19**

19a Burs fall apart easily into many separate achenes (they are all attached to a central RECEPTACLE).
Avens, p. 100, or White Avens, p. 97
19b Burs do not fall apart easily. . . . **20**

20a Burs are tiny, less than ¼″ long. . . . **21**
20b Burs are ½″ long or longer. . . . **22**

21a Burs taper at both ends, with several rows of bristles at each end.
Agrimony, p. 102
21b Burs are not as above; they are more egg-shaped and completely covered with bristles.   Enchanter's Nightshade, p. 225

22a Burs are oblong, with two horns at the tip; borne without stalks in the leaf axils.   Cocklebur, p. 186
22b Burs are round, with no horns; borne on stalks and at the ends of the branches.   Burdock, p. 184

23a Plant looks like a giant Dandelion; made up of many parachute-like silky hairs arranged in a sphere.   Goat's Beard, p. 180
23b Plant is not as above. . . . **24**

24a Stem of the plant is fuzzy and flimsy. Fruit consists of five small capsules attached at the top to a central STYLE but detached at the bottom and curling upward.   Cranesbill, p. 212
24b Plant does not have the above combination of characteristics. . . . **25**

25a Fruit stalks are in an UMBEL. . . . **26**
25b Fruit stalks are not in an umbel. . . . **28**

26a Fruit stalks are in a SIMPLE UMBEL. . . . **27**
26b Fruit stalks are in a COMPOUND UMBEL. Parsley family, p. 118

0   1   2   3   4

27a Plant is tiny (rarely more than 6″ high), usually with shiny ever-green leaves; fruits are BEAKED.   Goldthread, p. 82
27b Plant is not as above.   Lily family, p. 56

28a Fruit is a small ACHENE surrounded by three wings; fruits can be numerous, if present, and found in dense dangling clusters.
     Buckwheat family, p. 66
28b Plant does not have the above combination of character-istics. . . . 29

29a Fruit is a SILIQUE. Maybe all that remains is the silvery membrane that once separated the two halves of the fruit.
     Mustard family, p. 84
29b Fruit is not a silique. . . . 30

30a Plant is a fern, with many parallel stalks close together arising di-agonally from the main stem.   Fern family, p. 47
30b Plant is not a fern. . . . 31

31a Fruit, if present, is a small, knobby achene, but it is well hidden among many formless bracts. Branches arch upward and the whole plant is rather coarse.   Ragweed, p. 182
31b Plant does not have the above combination of character-istics. . . . 32

32a Stem is sharply triangular.   Water Plantain, p. 204
32b Stem is not sharply triangular. . . . 33

33a Stem is square, i.e., it has *four* distinct faces.* . . . 34
33b Stem is not square, although it may be sharply ridged. . . . 43

34a If you crush the flower heads, they smell good.   Mint family, p. 137
34b If you crush the flower heads, they have no smell. . . . 35

* A good way to count these faces without losing track is to put a pencil mark or a pin mark on each one as you go around the stem.

35a Fruits, or other remains, are on STALKS, long or short. . . . 36
35b Fruits, etc., are stalkless. . . . 40

36a Fruits are in tight clusters around the stem, on very short stalks.
                    Purple Loosestrife, p. 220, or Blue Vervain, p. 233
36b Plant does not have the above combination of character-
     istics. . . . 37

37a Flower remains consist of a CAPSULE that opens along most of its
     length into two main sections.
                    Figwort family, p. 154, or Mad-Dog Skullcap, p. 148
37b No such capsule is present. . . . 38

38a Stem bulges conspicuously at the NODES.   Pink family, p. 71
38b Stem does not bulge at the nodes. . . . 39

39a Dried calyx is star-shaped and all of the LOBES are roughly the
     same size. Fruits, if present, are small and round.
                                              Loosestrifes, p. 226
39b Dried calyx is BILATERALLY SYMMETRICAL and the lobes are
     of unequal size.   Mint family, p. 137

40a Dried calyces are in WHORLS around the stem.
                                              Mint family, p. 137
40b Dried calyces, or fruits, are in a SPIKE at the tip of the
     stem. . . . 41

41a Dried calyces are fuzzy; bottom lip of calyx is papery and fan-like.
                                              Selfheal, p. 138
41b Plant is not as above. . . . 42

42a Fruits are small (less than ¼″ long), too numerous to count, and
     enclosed by a small toothed calyx which opens at the top. They are
     oriented parallel to the stem.   Blue Vervain, p. 233
42b Fruits are about ½″ long, not too numerous to count, and oriented
     perpendicular to the stem.   Turtlehead, p. 158

| | | | | | | | | | | | | | | | | | | | | | | | | | | | | | | | | | | | | | | | | | | |
0   1   2   3   4

43a Fruits or flower remains are thorny; if you try to crush them in your hand, it hurts. . . . **44**

43b Fruits or flower remains are not thorny; not as above. . . . **47**

44a Fruit is a woody capsule divided into four sections, the sections separated by a lacy membrane.   Jimson Weed, p. 152

44b Plant is not as above. . . . **45**

45a Remains consist of many small cup-like bracts, arranged in rows along one side of the branches. Branches are curving and snake-like; whole plant is bristly-hairy.   Viper's Bugloss, p. 232

45b Plant does not have the above combination of character-istics. . . . **46**

46a Fruit heads are stiff and erect; subtended by narrow, curving, spiny bracts.   Teasel, p. 238

46b Fruit heads nod and consist of many overlapping thorny bracts.   Thistle, p. 188

47a Remains of flower is a RECEPTACLE conspicuously covered with fuzzy hairs. Perhaps all you will see is the fuzz. . . . **48**

47b Plant is not as above. . . . **50**

48a Receptacle is flat or rounded but not oblong.   Daisy family, p. 167

48b Receptacle is oblong, almost conical. . . . **49**

49a Flower stalks branch oppositely with a bulge at the node; hairs are white and woolly.   Thimbleweed, p. 80

49b Branching is alternate, forking; hairs are short and stiff.   White Avens, p. 97

50a Flowers remains are in a HEAD. . . . **51**

50b Flower remains are not in a head. . . . **54**

51a Head consists of fleshy bulblets that smell like onion, with perhaps a few flower stalks sticking out.   Field Garlic, p. 58

51b Head is not as above. . . . **52**

```
|!|!|!|!|!|!|!|!|!|!|!|!|!|!|!|!|!|!|!|!|!|!|!|!|!|!|!|!|!|!|!|!|!|!|!|!|!|!|!|
0           1           2           3           4
```

52a Heads are egg-shaped, brown, and fuzzy; usually found in the leaf axils along the stem. Bush-Clover, p. 105

52b Heads are not as above; usually found at the top of the stem. . . . **53**

53a Head is relatively long and thin, approximately the same width from top to bottom. Plantain, p. 236

53b Head is not as above. Daisy family, p. 167

54a Flower remains consist of a receptacle subtended or surrounded by several *overlapping* bracts. . . . **55**

54b Plant is not as above. . . . **59**

55a Plant is low and bushy, with dried flower remains usually lined up along upper side of branches; grows in salt marshes. Sea Lavender, p. 228

55b Plant does not have the above combination of characteristics. . . . **56**

56a Stems and stalks are slender, with a bulge at each NODE. Deptford Pink, p. 76

56b Stems do not bulge at the nodes. . . . **57**

57a Flower stalks are often, though not always, in WHORLS, and are oriented more or less perpendicular to the stem. Loosestrifes, p. 226

57b Plant does not have the above combination of characteristics. . . . **58**

58a Flower remains consist of cup-shaped dried bracts in two circles of five each, the outer bracts having a different shape or size from the inner. The bracts are often fuzzy and the branching is usually opposite. One plant usually has several to many flowers. Cinquefoil, pp. 95–96

58b Plant does not have all of the above combination of characteristics, even though it might have some. Daisy family, p. 167

59a Fruit is completely closed; has no opening whatsoever. . . . 60

59b Fruit, or other flower remains, has some kind of opening. . . . 66

0	1	2	3	4

60a Fruits are almost flat and stick easily to your clothes.

Tick Trefoil, p. 108

60b Fruits are not flat. . . . **61**

61a Fruits have no STALKS. . . . **62**
61b Fruits are borne on stalks. . . . **63**

62a Fruits are in a SPIKE; plant rarely grows more than 1 ft. tall

Plantain, p. 236

62b Fruits are in a head; plant grows up to 5 ft. tall.

Tall Meadow-Rue, p. 81

63a Plant rattles when you shake it.   Wild Indigo, p. 106
63b Plant makes no sound when you shake it. . . . **64**

64a Fruits are arranged in WHORLS or RACEMES along the
stem. . . . **65**
64b Fruits are only at the tips of the branches.   Pinweed, p. 218

65a Fruit stalks are more or less perpendicular to the stem; fruits less
than ¼″ in diameter.   Loosestrifes, p. 226
65b Fruit stalks arch upward; fruits about ½″ in diameter.

Moth Mullein, p. 156

66a Fruit, dried calyx, etc., has only one opening, at the top. . . . **67**
66b Fruit, dried calyx, etc., is not as above—either it opens lengthwise
along one or more seams, *or* it opens into more than one section. . . . **73**

67a Fruits, etc., are small, arranged in dense clusters around the stem.
Plant grows up to 4 ft. tall with branches like a candelabra.

Purple Loosestrife, p. 220

67b Plant does not have the above combination of
characteristics. . . . **68**

0   1   2   3   4

68a Fruits, etc., are in a SPIKE.   Plantain, p. 236
68b Fruits, etc., are not in a spike. . . . 69

69a Branching is opposite. . . . **70**
69b Branching is alternate. . . . **71**

70a Stems bulge at the NODES.   Pink family, p. 71
70b Stems do not bulge at the nodes.   Mint family, p. 137

71a Fruit is square.   Seedbox, p. 224
71b Fruit is not square. . . . **72**

72a Plant is low and bushy, with the dried flower remains usually lined
    up along one side of the stem; grows in salt marshes.
                                  Sea Lavender, p. 228
72b Plant is not as above.   Indian Tobacco, p. 241

73a Fruit, dried calyx, etc., opens lengthwise along one or two seams or
    into two main sections. . . . **74**
73b Fruit, dried calyx, etc., opens along more than two seams or into
    more than two main sections. . . . **79**

74a Fruit twists into a tight spiral after opening.   Bean family, p. 103
74b Fruit does not twist into a tight spiral after opening. . . . **75**

75a Fruits more than 1″ long. . . . **76**
75b Fruits less than 1″ long. . . . **77**

76a Fruits brown, 3″–4″ long; thin; dangling.   Dogbane, p. 230
76b Fruits are not as above.   Milkweed family, p. 130

77a Plant small, branches from the base; grows near Beech trees. Each
    fruit stalk is subtended at the AXIL by a single pointed bract.
                                  Beech-drops, p. 234
77b Plant does not have the above combination of character-
    istics. . . . **78**

78a Plant has many loose branches that rattle when you shake them.

Wild Indigo, p. 106

78b Plant does not have loose branches that rattle.

Figwort family, p. 154

79a Fruit, etc., opens along three seams or into three main
sections. . . . **80**

79b Fruit, etc., opens along four or more seams. . . . **87**

80a Fruits are closed at each end and open only by slits along the side.

Orchid family, p. 63

80b Fruits, etc., are not as above. . . . **81**

81a Fruits are more than 1″ long.   Iris family, Lily family, p. 56
81b Fruits are less than ½″ long. . . . **82**

82a Stem appears flat; actually has two wings running its full length.

Blue-eyed Grass, p. 62

82b Stem is not as above. . . . **83**

83a Fruits are in a SPIKE; plants small.   Sundew, p. 209
83b Fruits not in a spike. . . . **84**

84a Fruits, dried SEPALS, etc., are in WHORLS or RACEMES with the
STALKS roughly perpendicular to the main STEM.

Loosestrifes, p. 226

84b Fruits, etc., are found at the tips of the branches or in PANICLES
along the stem. . . . **85**

85a Most branching is opposite.   St. John's-wort, Pineweed, p. 216
85b Most branching is alternate. . . . **86**

86a Fruits are tiny, mainly borne at the tips of the many delicate
branches; plant grows in dry soil.   Pinweed, p. 218

86b Fruits are not tiny; are borne in panicles along the stem; plant
grows in standing water.   Water Willow, p. 222

87a Fruits, dried calyces, etc., are tubular—longer than they are wide. . . . **88**

87b Fruits, etc., are round—wider or as wide as they are long. . . . **93**

88a Fruits are in a SPIKE. . . . **89**

88b Fruits are not in a spike. . . . **90**

89a Stem and fruits are closely covered with a flannel-like fuzz.
Common Mullein, p. 155

89b Stem and fruits are smooth or perhaps slightly hairy, but not fuzzy.
Evening Primrose, p. 223

90a Fruits are borne only at the tips of the branches. . . . **91**

90b Fruits, dried calyces, etc., are borne all along the branches. . . . **92**

91a Branching is opposite, with a bulge at each NODE.
Pink family, p. 71

91b Branching is alternate, with no bulge.　　Columbine, p. 79

92a Whole plant is bristly-hairy, almost thorny.
Viper's Bugloss, p. 232

92b Plant is not bristly-hairy.　　Sea Lavender, p. 228

93a Fruit is a three- to six-parted capsule wrapped in a frilly calyx; arranged in panicles. Plant grows in standing water.
Water Willow, p. 222

93b Plant is not as above. . . . **94**

94a Fruits or fruit heads are 1″ or more in diameter.
Mallow family, p. 115

94b Fruits, dried calyces, etc., are ½″ or less in diameter. . . . **95**

95a Fruits are borne on the upper side of branches that arch from the top of the stem.　　Ditch-Stonecrop, p. 210

95b Fruits are not as above. . . . **96**

96a Flower remains are flattened, star-like SEPALS, borne on stalks more or less perpendicular to the stem.   Loosestrifes, p. 226
96b Plant is not as above. . . . 97

97a Branching of flower stalks is opposite. . . . **98**
97b Branching of flower stalks is alternate, or plant has no branches. . . . 99

98a Flower remains consist of a tubular, toothed calyx which is bilaterally symmetrical. Flower stalks are all along the stem.
                                                                    Mint family, p. 137
98b Flower remains consist of a cup-shaped, RADIALLY SYMMETRICAL dried calyx, and the flower stalks are all at the top of the stem.   Cinquefoil, pp. 95–96

99a Plant grows 1 ft. or more tall, in open places. Fruits are tiny and too numerous to count.   Meadowsweet, Hardhack, p. 98
99b Plant grows less than 1 ft. high, usually in the woods, with one or few fruits on each plant.   Wintergreen family, p. 126

100a Flower stalks are almost, but not quite, in WHORLS around the stem.   Joe-Pye Weed, p. 189
100b Flower stalks are not in whorls. . . . **101**

101a Flower stalks are in an UMBEL.   Parsley family, p. 118
101b Flower stalks are not in an umbel. . . . **102**

102a Plant is a fern, with many dried leaves rising diagonally from the stem parallel to each other.   Fern family, p. 47
102b Plant is not a fern. . . . **103**

103a Flower stalks look like little pegs oriented more or less perpendicular to the stem.   Garlic Mustard, p. 89
103b Plant not as above. . . . **104**

104a Stem is stout and bamboo-like, with leaf scar wrapping around it at each NODE. Plant grows up to 9 ft. tall, often in clusters. Only remains are clusters of zig-zag flower stalks.

Japanese Knotweed, p. 67

104b Plant does not have the above combination of character-istics. . . . **105**

105a Stem is coarse and bristly. Great Ragweed, p. 182

105b Stem is not coarse and bristly. . . . **106**

106a Leaf scar forms a slight bulge atthe nodes.

Tall Meadow-Rue, p. 81, or Blue Cohosh, p. 208

106b Leaf scar does not bulge at the nodes; plant is totally nondescript.

Pigweed, p. 207, or Common Ragweed, p. 182

GLOSSARY

achene A small, dry fruit that does not open. It contains one seed and looks like a seed itself.

alternate Not opposite; borne at different levels on the stem. Applies to leaves, stalks, or branches.

annual A plant which dies, roots and all, at the end of the growing season. An annual depends on dissemination and germination of its seed to continue its existence.

axil The upper angle between a leaf, or its stalk, and the stem (from Latin *axilla*, "armpit").

basal (leaves) At the base of the plant.

beak A long, pointed tip.

biennial A plant with a two-year life cycle which produces leaves in its first year and flowers in its second.

bilaterally symmetrical Symmetrical along one axis, like a person.

brackish (water) Containing salt, but in a lower concentration than sea water.

bract A leaf-like structure that you sometimes find under or surrounding the flower, similar to but often larger than the *sepals*. The bracts often dry and remain through the winter.

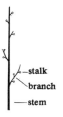

branch A division of the stem, different from a *stalk*.

—stalk
branch
—stem

bur A fruit, cluster of fruits, or fruit covering that detaches easily from the plant and sticks to clothes and fur by means of hooks or hairs.

(Burdock)

calyces The plural of *calyx*.

calyx The sepals, collectively; the outer circle of floral leaves. If the sepals are fused together, the whole structure is referred to as the *calyx*.

— corolla
— calyx
(Evening Lychnis)

capsule A dry fruit, usually rounded, usually opening into more than one section. The number of sections is useful in identification.

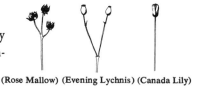

(Rose Mallow) (Evening Lychnis) (Canada Lily)

compound (leaf) Made up of separate, smaller leaflets.

compound leaves

compound umbel An umbel in which a second umbel emanates from the tip of each primary flower stalk.

(Queen Anne's Lace)

corolla The petals collectively; the inner circle of floral leaves.

— corolla
— calyx
(Evening Lychnis)

family A broad grouping of plants, based mainly on flower characteristics.

Glossary **37**

fruit A ripened ovary; the part of the plant that contains the seeds.

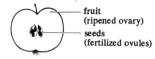

fruit
(ripened ovary)
seeds
(fertilized ovules)

genus A group of plants within a family; a general category such as Goldenrod, or Lily. A genus is a more general category than *species*.

germinate To sprout; to begin to grow from a seed.

hairy Having any hairs whatsoever: fuzzy, downy, woolly, cottony, silky, flannelly.

head A tight cluster of stalkless or almost stalkless flowers or fruits at the tip of the stem.

(Bush-Clover)

herbaceous Not surviving above ground over the winter; non-woody.

humus One of the upper layers of soil, made up of decayed or decaying organic matter—dead leaves, dead animals, etc.

inflorescence A cluster of flowers.

legume The characteristic fruit of the Bean family; also refers to any plant that is in the Bean family. A legume is more or less synonomous with a pod: usually a long bilaterally symmetrical dry fruit that opens its full length along two seams.

closed

open

lobe A division.

node A point on the stem where branches or leaves originate.

node

oblong Longer than wide, with sides nearly parallel.

opposite Paired; borne at the same level on the stem. Applies to leaves, stalks, and branches.

ovary The part of the flower, usually hard to see, that contains the ovules.

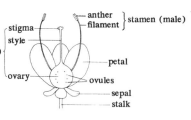

ovule A multicellular structure which contains the female reproductive cells. After fertilization, the ovule becomes a seed.

panicle A compound raceme; an inflorescence where the flowers are borne on stalks that branch off larger stalks.

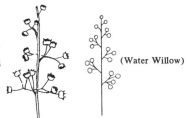

(Water Willow)

pappus In the Daisy family, fine hairs or bristles attached to the achene. Technically, the pappus is part of the calyx.

pappus

parasite A plant or animal which obtains all of the elements necessary for its survival from another living organism and does not contribute anything to the survival of its host.

perennial A herbaceous plant which dies only to the ground at the end of the growing season. The underground parts stay alive and produce new shoots the following spring.

pistil The female part of the flower, consisting of the stigma, the style, and the ovary.

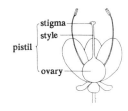

pod A dry, usually elongated fruit that opens its full length along one or two seams.

(Common Milkweed)

prostrate Growing horizontally along the ground.

raceme An arrangement of stalked flowers along a central stem, usually with space between each flower.

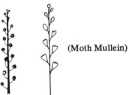
(Moth Mullein)

radially symmetrical Symmetrical along many axes, like a wheel.

receptacle The part of the flower to which the stamens, pistils, petals, fruits, and other parts are attached. In many species, the receptacle is merely the tip of the flower stalk, but in some it is expanded to various shapes. This is especially true for many species in the Daisy family, where the receptacle is the base to which all of the composite flowers are attached.

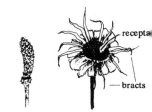
receptacle
bracts
(Thimbleweed) (Elecampane)

rhizome An underground stem which crawls along and sends out leafy shoots from its upper side and roots from its lower.

root The part of the plant which draws water and nutrients out of the soil into the plant.

rootstock Rhizome.

rosette A cluster of leaves, usually basal, in a circular pattern.

salt marsh A wet area near the shore flooded daily by salt water. Salt marshes are characterized by certain plant species.

saprophyte A plant that obtains all its nourishment from dead organic matter.

seed A fertilized ovule. The seed contains the embryo, which becomes the new plant.

sepals Small leaf-like structures that sometimes surround the flower; usually green, in contrast to the usually brightly colored petals.

petals
sepals
bract

silique A two-parted fruit divided down the middle by a membrane. In several species the membrane is silvery and persistent even after the outer walls of the fruit have disintegrated and the seeds have been dispersed. A silique is the characteristic fruit of the Mustard family.

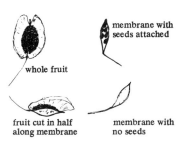

whole fruit

membrane with seeds attached

fruit cut in half along membrane

membrane with no seeds

simple umbel An umbel with only one set of flower stalks.

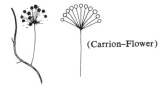

(Carrion–Flower)

species A particular kind of plant within a genus, such as Grass-leaved Goldenrod as opposed to Seaside Goldenrod.

spike A dense cluster of stalkless or almost stalkless flowers (or fruits) along a central stem.

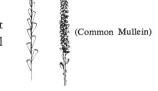

(Common Mullein)

stalk A stem-like structure that supports a leaf, fruit, or flowers; different from a *branch*.

stem The main axis of the plant, through which nutrients and water pass between the leaves and the roots; usually serves as support, except for a vine or a prostrate plant.

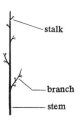

stalk

branch
stem

stigma The top part of the pistil, often sticky, where pollen lands.

see drawing below

stipule

stipule A tiny leaf at the base of the leaf-stalk.

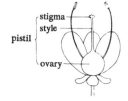

style The elongated part of the pistil that connects the stigma and the ovary.

pistil — stigma / style / ovary

subtend To occur below.

tendril A slender branch-like structure which curls about and grabs onto other objects; usually found on vines.

tendril

thorn A sharp, pointed protuberance.

umbel An arrangement of flowers (or fruits) where all the flower stalks radiate from one point at the tip of the stem (from Latin *umbella,* "umbrella").

(Carrion–Flower)

waste place An area with poor soil created by human activity. Waste places include vacant lots, roadsides, railroad embankments, filled areas, old gravel pits, and the like.

whorl A circle of stalks or leaves *around* the stem, all emanating at the same level.

(Whorled Loosestrife)

wraparound (leaves) Encircling the stem at the base of the leaf.

SOURCES AND
RECOMMENDED READING

If you want to see what these plants look like in bloom, there are scores of wildflower books available, in all sizes and prices. I recommend these:

A Field Guide to Wildflowers, by Roger Tory Peterson and Margaret McKenny (Houghton Mifflin, 1968). By far the best for identification. Mainly black and white drawings, some colored. Inexpensive, and available in almost all bookstores.

Wildflowers of the United States: The Northeastern States, by Harold N. Rickett (McGraw-Hill, 1966). Two volumes; expensive; beautiful color photographs.

A Field Guide to the Ferns, by Boughton Cobb (Houghton Mifflin, 1963).

You might find others more suitable to your taste and purse. Look around.
   For professional technical botany, there are two authorities:

Gray's Manual of Botany, 8th ed., by Merritt Lyndon Fernald (D. Van Nostrand, 1950). Very technical; illustrations too small to be useful. One volume.

The New Britton and Brown Illustrated Flora of the Northeastern United States and Adjacent Canada, 3rd revised ed., by Henry A. Gleason (Hafner, 1968). Three volumes. The main advantage of Britton and Brown over Gray's is that it has a detailed line drawing of every species identified.

Both of these are expensive and will probably have to be ordered specially by your bookstore. Most college libraries should have both.
   To identify other winter specimens (trees and insect galls), the best books to start with are:

Fruit Key and Twig Key to Trees and Shrubs, by William Harlow (Dover, 1946).

Woody Plants in Winter, by Earl Core and Nelle Ammons (Boxwood, 1958).

Plant Galls and Gall Makers by E. P. Felt (Hafner, 1965).

If you want to know more about eating wild plants, this is a very popular subject on which many books have recently been published. The classic is *Stalking the Wild Asparagus,* by Euell Gibbons (David McKay, 1962).

If you want to know what *not* to eat, try *Poisonous Harvest,* by John M. Kingsbury (Holt, Rinehart and Winston, 1965), or *Poisonous Plants of the United States and Canada,* by John M. Kingsbury (Prentice-Hall, 1964).

THE PLANTS

FERN FAMILY
POLYPODIACEAE

Ferns do not produce flowers and fruit in the same way as the rest of the plants in this book. They reproduce by means of spores—dust-like particles which are contained in spore cases on the plant. Generally the spore cases are on the underside of the normal leaves; sometimes, as is the case with the two ferns in this book, they are borne on separate specialized leaves. When the time is right, the spore cases break open and the spores are dispersed. They germinate into a small, flat, leaf-like structure on the underside of which sexual organs eventually develop. The male sexual organ produces sperm which swim to and fertilize the egg in the female organ. From this, a new fern as we know it develops.

Ferns have a characteristic herringbone leaf structure. They are all perennials and most of them grow in wet, shady places.

If you want to see what your ferns look like in summer, or to identify more ferns, look at A Field Guide to the Ferns, *by Boughton Cobb (Houghton Mifflin, 1963).*

before spore cases have opened

after spore cases have opened

detail of stem

Ostrich Fern
Matteuccia Struthiopteris

The general shape of this fern, both in leaf
and in fruit, is supposedly like an ostrich
feather—widest towards the top. The fruiting
stalks grow about one to two feet high and
have a deep groove down one side of the
axis. The spore cases are arranged in long,
thin, rounded "pods" along most of the stalk.
You have probably noticed Ostrich Fern in
the summer; its leaves grow in a basket as
high as five feet and it is quite impressive.
Like other ferns, Ostrich Fern grows in moist
shady places.

after spore cases have opened

before spore cases have opened

Sensitive Fern
Onoclea sensibilis

Sensitive Fern is easy to recognize by its bead-like fruiting structures. If you break open one of these beads, you will find many hard, round spore cases. Break open a spore case on a piece of white paper, and if you have good eyes you will see the dust-like spores. After the spore cases have opened on the plant, the beads no longer look like beads, but they are lacy and look somewhat like an open hand. Sensitive Fern is one of the few plants in this book with a definite front and back; the spore cases are all lined up on one side of the plant. The fruiting stalks of Sensitive Fern grow about one foot tall, in wet places.

GRASS FAMILY
GRAMINEAE

Economically, the Grass family is one of our most important. Wheat, barley, oats, rye, sorghum, millet, corn, and sugar cane are among its members. Botanically, it is one of the hardest families to describe and its members are among the hardest to distinguish. Fortunately, most people have a good instinctive sense of what is a grass and what is not a grass.

The flowers in the Grass family are inconspicuous and very much reduced—they do not have petals and sepals like most of the other flowers in this book. Flower characteristics are not much help in winter, however, but grass stems are also characteristic. They are brittle, straw-like, and hollow at the nodes. Also, most grasses have relatively narrow, pointed leaves which wrap completely around the stem down to the next node. If you try to take a leaf off the plant, you will end up taking with it the whole outer layer of the stem below it.

The fruit in the Grass family is a grain, normally too small to help you in identification.

Orchard Grass
Dactylis glomerata

This grass grows in every part of North America except for the desert and the Arctic. In our area, it sets seed and starts to dry out by mid-summer. It reaches a height of four feet and is found in fields and meadows, and along roadsides.

Reed Grass
Phragmites communis

Phragmites is a plant of industrial America. If you go south from New York City, by train or by car, you first pass through thousands of acres of *Phragmites,* dotted with an occasional billboard or radio tower. With the recent environmentalist concern about salt marshes, *Phragmites* has become the villain—the symbol of a dead salt marsh which no longer supports salt marsh vegetation. *Phragmites* can become established in a salt marsh if the daily flow of salt water over the marsh is interrupted. This can happen if a tide gate is built at the mouth of the marsh, restricting the entry of salt water, or if the marsh is filled with dredging spoils which raise its elevation above the high-tide mark. Once *Phragmites* becomes established, it takes over in a hurry, by many methods. In stands that are not very dense, shoots will fall over and become horizontal runners. These send up new shoots from the nodes, and they put out new rhizomes, underground stems which anchor the plant and which also creep along and send up new shoots. *Phragmites* catches fire easily, but its rhizomes are buried deep in the mud, so they are rarely injured, and they send out new shoots after the fire, often in a denser growth than before. Strangely enough, though, in spite of the fact that it covers so much of the eastern seaboard, very little is known about the ecology of *Phragmites*—what limits its distribution and how or why to control it. It would be logical to try to find a use for *Phragmites*: what other crop grows so well with so little care, and so close to the major centers of population? In England, the plant has been used for thatching, and areas are often manipulated, for example, by fire, to increase their yield. In this country, however, scientists have only recently begun to investigate the possibilities. The most promising possibility seems to be the use of *Phragmites* as a source of cellulose.

With a maximum height of about 20 feet, *Phragmites* is one of the tallest herbaceous plants in the Northeast. You can recognize it by its height, its brittle, bamboo-like stems, and its fluffy gray seed heads. *Phragmites* is found world-wide.

If you want to collect *Phragmites* for decoration, be sure to get it early in the fall—say September—while the seed heads still have some body. If you wait too long, the seeds will be dispersed, the rains will beat down on the plant, and you will have nothing left except scraggly brown stalks. Once you collect *Phragmites,* you do not need to worry about its "exploding" like a Cat-tail (p. 200).

Although *Phragmites* has various common names, it is most often referred to simply as *Phragmites.*

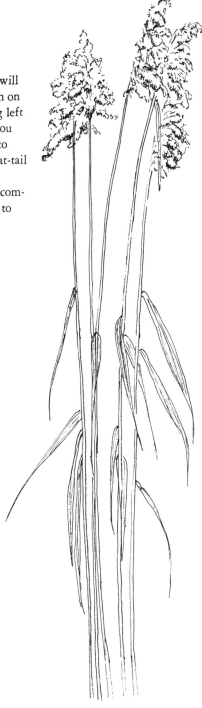

Love Grass
Eragrostis spectabilis

(Tickle Grass, Petticoat Climber)

This is a short grass that rarely grows more than a foot high. Unlike most grasses, the flowering stalks arise almost at ground level. They form a delicate panicle which is widest at the middle and tapers at both ends. In winter, the panicle breaks off from the stem and you usually find it floating along on the ground like a tumbleweed. The spikelets are a distinctive reddish-purple. Notice the tufts of hair in the axils of the flowering stalks. This species grows in dry soil along roadsides, in fields and in open woods.

Foxtail Grass
Setaria glauca

You can recognize this grass easily by its fuzzy tufts of tan or yellow hairs, at the base of which you might find a few grains. This is in the same genus as the cultivated Millet, *Setaria italica.* Our species grow as high as two and a half feet, but you often find it growing in poor soil where it rarely exceeds a foot. Aside from growing in waste places, Foxtail Grass is also a common garden weed.

LILY FAMILY / LILIACEAE
IRIS FAMILY / IRIDACEAE

Although these two families are different when the plants are in bloom, the differences do not show enough in winter to make the distinction worthwhile. They are both part of a large group of flowering plants called monocots. *One of the main characteristics of the monocots is that the flowering parts are in multiples of three. This means that the fruits are also divided into threes, and with one exception on the following pages (Smilax), you will see three- or six-parted capsules. Another characteristic of the monocots is that the veins on their leaves run parallel to each other, not in a feather pattern like those of many other plants. Some of these dried plants might still have leaves on them, and you can see this characteristic venation.*

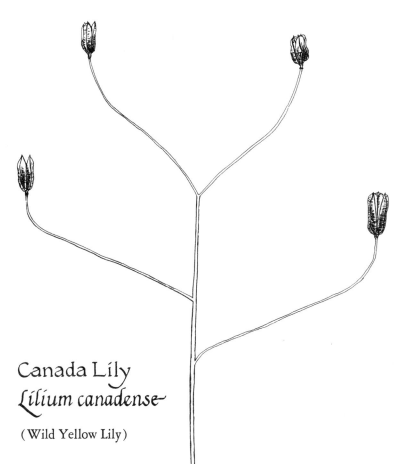

Canada Lily
Lilium canadense

(Wild Yellow Lily)

You might find this plant growing as a single stem, but more often it branches like a candelabra. Its fruit is a three-parted oblong capsule which opens only slightly. Inside the capsule you will perhaps find flat, papery seeds. The leaves are borne in whorls around the stem, as you can see from the whorled leaf scars. Canada Lily grows quite tall—up to six feet—in meadows, woods, thickets, and clearings. It is a perennial, growing from a bulb. Although its range extends through the Northeast, it is most common in northern areas.

Field Garlic
Allium vineale

This is a common weed in lawns and meadows,
growing from a small onion-like bulb. Its hollow stems
grow as high as three and a half feet and produce a
cluster of little bulbs at the tip. Each bulb has a short
curving tail. Some plants produce only bulbs, but some
produce a few flowers, sticking out on short stalks
above the bulblets. The flowers become three-parted
capsules, and you might find a few still on the plant.
The cluster of flowers and/or bulblets is always subtended by a short pointed bract.

Although Field Garlic is in the same genus as the
garden onion (*Allium Cepa*), it is considered by
most people to be too strong to eat. Dairy farmers
especially do not like it, since the cows often eat it,
and it flavors their milk.

Wild Leek
Allium tricoccum

This plant grows about six inches to one foot tall, in rich woods. If you pull it up hard enough when you pick it, you will notice its most obvious feature— the strong onion-smelling bulb. Wild Leek usually grows in extensive patches, and the early spring woods sometimes reek of it. If you don't notice the bulb, you can still recognize the plant by the large black seeds borne in an umbel. These seeds are contained inside three-parted capsules, but the capsule walls open downwards under the seeds, so that you might first think they were sepals. If you have ever grown onions, you will see the similarity between the fruits in your garden and those of Wild Leek. Both are in the genus *Allium;* and the bulbs of the Wild Leek are perfectly safe to eat. Euell Gibbons (*Stalking the Wild Asparagus*) finds them "the sweetest and the best of the wild onions." They are also one of the largest, which makes your collecting and preparation time a little more worthwhile.

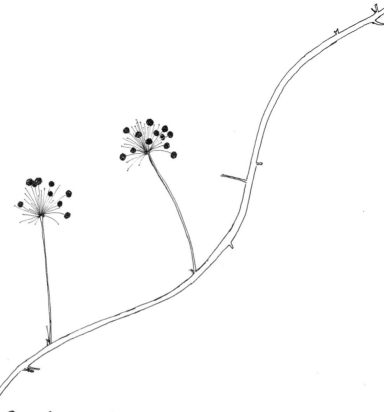

Carrion ~ Flower
Smilax herbacea

You might know the genus *Smilax* as bullbrier, catbrier, greenbrier, or horsebrier—a very unpleasant bramble that makes large areas of woods impenetrable. Carrion-flower, however, is not so offensive as the others; it is one of the few species of *Smilax* without thorns, and it does not grow in a tangled mass like the other members of the genus. It usually grows as a single stem, up to six feet high, and clings to other plants with its tendrils, but these disappear in winter. Its fruit is a blue berry borne in an umbel. If the berries have disappeared, you can still recognize the plant, for you will still see the flower stalks in a characteristic umbel. There is no other plant in this book with an umbel which also grows as a vine.

The English name refers to the smell of the flower.

seed

Blue Flag
Iris versicolor

This plant grows two to three feet tall, and the capsule is one and a half to two inches long. It often stays closed till late in the winter, but when it opens it flares out like a blossoming flower. The capsule is divided into three main sections, each one of which is again divided in half. The seeds are also three-sided, and you will probably find some stacked up inside each section of the capsule. Although the fruits of the Iris and the Lily are somewhat similar, you can identify the Iris by its wraparound leaves. You probably know the flat, fan-like leaves of an Iris in spring, and you might find remains of these leaves subtending the flower. Even if the leaf itself is not there, you will see its wraparound scar. The Lily's leaves are in whorls, and they do not wrap around.

Iris capsules are a rare find, even though Iris is not an uncommon flower. The capsules are also fragile, and it is hard to find one intact. Iris grows in wet places from a spreading underground rhizome.

Blue-eyed Grass
Sisyrinchium spp.

This tiny plant will escape your notice unless you look for it, for it usually grows in among the grass in meadows, woods, and clearings. Aside from the small size of its fruits, all clustered at the top of the stem, you can recognize the plant most easily by its distinctive stem. From a distance, the whole stem looks flat, but up close, you realize that it is a conventional round stem with two wings running its complete length. The plant rarely grows more than one foot tall, and the capsules are usually less than a quarter of an inch wide. The capsules open into three main sections.

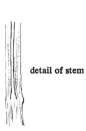

detail of stem

ORCHID FAMILY
ORCHIDACEAE

*To many people's surprise, there are several species of orchids grow-
ing wild in the Northeast. Some of them, like Lady's-Slipper, are
fairly common, while others are quite rare. Most of our native
orchids have much smaller and more delicate flowers than the garish
sprays you pay such a high price for at the florist's.*

*The fruit of the orchids is a three-parted capsule, but the fruits of
most of them stay closed at each end and open by slits along the sides.
Their seeds are minute, like dust.*

*Orchids branch very little. The flowers are arranged singly, in a
spike, or in a raceme. Many orchids grow in bogs and other places
with acid soils.*

Rattlesnake Plantain
Goodyera pubescens
(Lattice Leaf)

When you find this perennial in the woods, you can recognize it most easily by its distinctive basal leaves, which last through the winter. They are dark green and covered with a net-like pattern of white veins. The flower stalks are usually six to eight inches tall, and the papery capsules are arranged in a spike. The capsules look like inflated sacs. Like those of other orchids, they are closed at each end but open by slits along the side. They are fairly small, less than a quarter of an inch long. Rattlesnake Plantain grows in dry or moist woods, often nestled in between rocks.

Lady's ~ Slipper
Cypripedium acaule

(Moccasin Flower)

This plant grows up to one foot tall, and there
is only one capsule on each plant. The capsule is
woody and about one or two inches long. It is
closed at both ends but opens along six slits. It
also has six lengthwise ribs, and late in the winter you might find it all disintegrated except for
these ribs. Notice the long curved bract at the
base of the capsule. The Lady's-Slipper flower
stalk rises up from a pair of large basal leaves
which completely disappear by winter, so you
will not see any leaf scars on the stalk.

Lady's-Slipper likes acid soil. You often find large patches
of it growing in pine woods. In the northern part of its
range, towards Newfoundland, you also find it growing in
bogs and wet woods. South New Jersey and in northern
Indiana, Lady's-Slipper confines itself to the mountains and
the coastal plain.

Lady's-Slipper capsules are a rare find. You can go to a
place that is full of Lady's-Slippers in the spring and not
find a single capsule in the winter. The reasons for this are
not clear. If a flower is not producing fruit, an obvious possible reason is that the flower is not getting pollinated. Some
flowers are built so that they can only be pollinated by certain insects, and if those insects are rare, the flower might
not get pollinated very often. However, Lady's-Slipper
seems to be pollinated by several species of several genera—
bumble bees (genus *Bombus*), miner bees (genus *Andrena*),
and leaf-cutter bees (genus *Megachile*). None of these
genera are particularly rare in the Northeast; maybe they
just do not visit the Lady's-Slipper, for some reason or other.
This question needs more exploration.

BUCKWHEAT FAMILY
POLYGONACEAE

Although we describe only two members of the Buckwheat family, it has many common members which you are likely to find in your wanderings. It has two obvious winter characteristics: wraparound leaf scars and three-winged fruits. In summer, the base of the leaves encircles the stem and leaves a wraparound scar when it falls off. It sometimes leaves a slight bulge. The fruit is an achene, usually sharply triangular. Most of the flowers have three sepals which enlarge as the fruit ripens and form wings around it. Sometimes the sepals form a three-winged sac that encloses the fruit; sometimes the wings sit underneath the fruit. The fruit and wings are all small—usually less than a quarter of an inch long.

Two commercial species in the Buckwheat family are buckwheat itself (Fagopyrum esculentum) *and rhubarb* (Rheum Rhaponticum). *You might object that rhubarb does not have wraparound leaves, but you simply cannot see the sheathing, because the leaves rise straight from the ground. (The red stalk that you eat is part of the leaf; it is not the stem.) The garden books tell you not to let your rhubarb go to flower because the flowering stalk will sap nutrients from the root that would otherwise produce big stalks for you to eat next year. This is true, but if you have a lot of plants, you should let a few of them flower for your botanical education. You will see some wraparound leaves on the flowering stalk, and you will also see the characteristic flowers and three-winged fruits.*

Japanese Knotweed
Polygonum cuspidatum

(Mexican Bamboo, Wild Rhubarb)

This plant is huge. It grows as high as nine feet tall, usually in clumps. Ironically, it was introduced to this country to be planted for ornament, but it is now such an aggressive weed that a proper gardener would die if it were seen in his or her garden. The plant spreads by underground rhizomes, so if you want to dig up one you must dig up the whole clump, and you might end up digging up half your backyard. It is a perennial, so if you leave some rhizomes, the plant will come up again next year and start spreading all over again.

To its credit, Japanese Knotweed is good to eat in the early spring. You have probably noticed the shoots coming up. They are green and red spotted, and the leaves are still folded around the stem, so that the shoots look like little daggers poking through the sand. If you cut up the young shoots and cook them with sugar, they taste like rhubarb. Once you see the plants coming up, do not wait around for a few days before you pick them, because they grow frighteningly fast, and if they are any higher than eight inches or so they will be too stringy to eat.

In winter, the grooved stems of the Japanese Knotweed turn reddish-brown, and they are hollow except at the nodes. The stems do not grow straight, but arch out at the top, and the flower stalks zig-zag up into the air. The fruits usually do not stay long on the plant, but you might find some, and you will see that they are characteristically three-angled, with three wings.

Japanese Knotweed grows just about anywhere—vacant lots, backyards, roadsides. Unlike many of our waste place weeds, it is just as happy in moist sites as dry, and you often find it along stream banks or in wet spots.

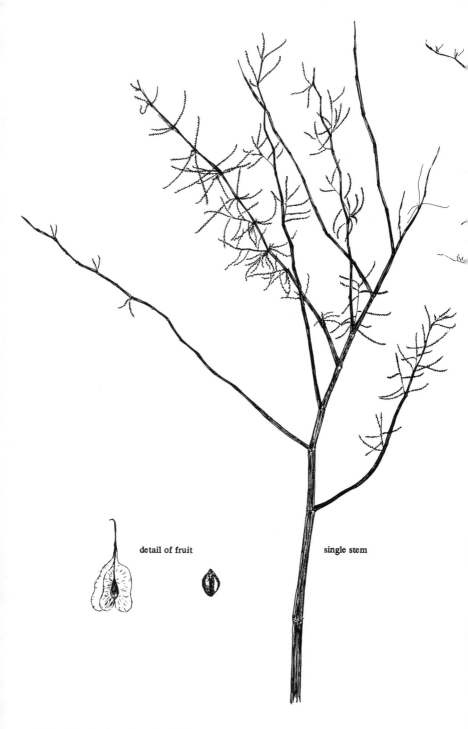

detail of fruit   single stem

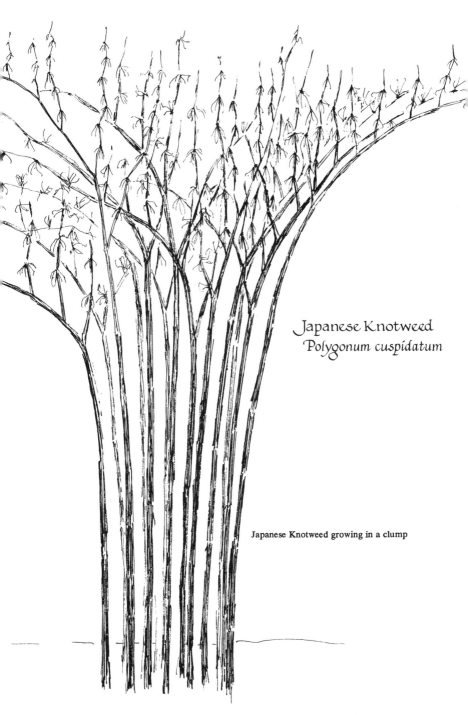

Japanese Knotweed
Polygonum cuspidatum

Japanese Knotweed growing in a clump

three-winged fruit

Dock
Rumex crispus

Dock grows two to three feet tall, in fields and waste places. It flowers and sets fruit in the spring, and you can see its straight, brown stalks growing among the grass and the Queen Anne's Lace by the middle of July. You can recognize Dock by its dried three-winged, heart-shaped sepals, which hang in dense umbrella-like clusters from the stems. Tucked in between each wing you might find a shiny three-angled fruit. As is characteristic of the Buckwheat family, Dock has wraparound leaf scars.

There are many species of Dock. In general form, most of them look like this one, but their fruits and winged sepals are all shaped slightly differently and some are quite intricate. They are easy to distinguish if you have a good picture book; the best source is *Britton and Brown.* If you have no access to a copy of *Britton and Brown,* however, Peterson's wildflower guide will help you for the more common species.

PINK FAMILY
CARYOPHYLLACEAE

The characteristics of this family are easy to spot. All of the plants in the Pink family have opposite branching and the conspicuous bulge in the stem at each node. (When you are looking for opposite branching, look carefully for stubs of broken branches. Often one branch of a pair will be broken off, or maybe never have fully developed, so that at first glance you might think the branching is alternate.)

In flower, many members of the Pink family have a toothed or notched calyx which dries and often completely surrounds the fruit. The calyx often has prominent veins, the number of which can help in identification. The fruit is a one-celled capsule, often toothed at its opening. The number of teeth is consistent with certain genera and can also be a useful feature in identification.

The most well-known cultivated plant in the Pink family is the carnation. Look at your next carnation for sale in the New York subway and you will see the characteristic opposite leaves and bulging stem.

Evening Lychnis
Lychnis alba

(White Campion)

Evening Lychnis goes through several stages in the development of its fruit. While the plant is in flower, it has a tubular, toothed, very hairy calyx. After the flower is fertilized, and the fruit begins to mature, the calyx inflates and completely covers the fruit. As the winter wears on, the calyx disintegrates, leaving the smooth, shiny capsule. You will probably find fruits with the calyx and without it. The opening of the capsule is usually surrounded by ten teeth, occasionally eight, and very occasionally six. Except for the capsule, the whole plant is hairy.

If you have a similar fruit with only six teeth, it could be a very similar species, *Silene noctiflora* (Night-flowering Catchfly). The only way to be sure which is which is to count the veins on the calyx; *Lychnis* will have twenty, *Silene* only ten.

Both plants grow in fields and waste places, and both flower in the evening or the night, as their names indicate. *Lychnis alba* can grow as high as four feet, but it rarely stands up straight. You usually find it sprawling among the other plants it is growing with.

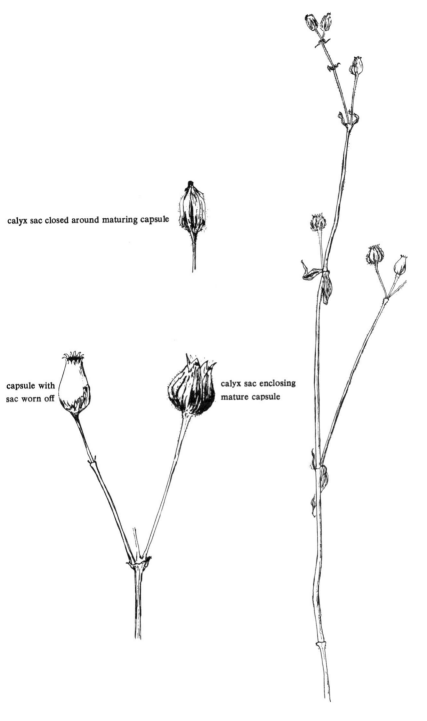

calyx sac closed around maturing capsule

capsule with
sac worn off

calyx sac enclosing
mature capsule

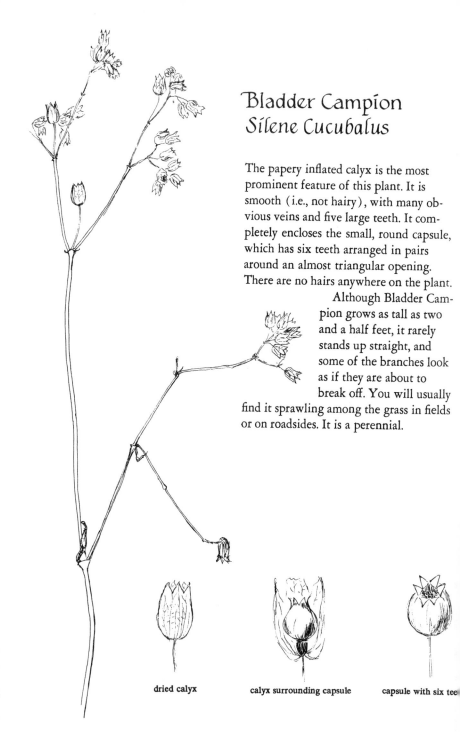

Bladder Campion
Silene Cucubalus

The papery inflated calyx is the most prominent feature of this plant. It is smooth (i.e., not hairy), with many obvious veins and five large teeth. It completely encloses the small, round capsule, which has six teeth arranged in pairs around an almost triangular opening. There are no hairs anywhere on the plant.

Although Bladder Campion grows as tall as two and a half feet, it rarely stands up straight, and some of the branches look as if they are about to break off. You will usually find it sprawling among the grass in fields or on roadsides. It is a perennial.

dried calyx　　　**calyx surrounding capsule**　　　**capsule with six tee**

Sleepy Catchfly
Silene antirrhina

Most specimens of this plant will attract your
attention by their sticky stems. The sticky substance
is concentrated in zones between the nodes. Insects
are said to get trapped there, which explains the
origin of the English name, "Catchfly." The Latin
name *Silene* is from the Greek word *sialon,* "sa-
liva," in reference to the sticky stems. (The reason
it is called "Sleepy" Catchfly is not clear.)

Not all of the plants are sticky, so if you find a
smooth one you can still recognize it by the opposite branching and the tiny capsules—one-tenth to
one-half inch long. Around the opening of each
capsule are six teeth. The capsules are completely
covered by the calyx, but it fits so closely that it is
hard to tell which is which.

Sleepy Catchfly grows in dry, sandy soil, eight
inches to two and a half feet tall. It is sometimes an
annual, sometimes a biennial.

detail of capsule

Deptford Pink
Dianthus Armeria

This is a small, delicate biennial often found on roadsides and in other places with poor, dry soil. As is
characteristic of the Pink family, it has opposite
branching and a bulge at each node. The fruit is a
slender capsule with four teeth at its opening, but it
is fragile, and it is hard to see because it is almost
completely surrounded by the calyx and by many
bracts. The calyx has long, thin, strongly curled lobes,
one of the most distinctive features of the plant. The
fruits are borne opposite each other in tight clusters.
It can grow to a maximum height of a foot and a half,
but it is usually less than a foot tall.

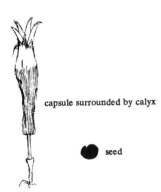

capsule surrounded by calyx

● seed

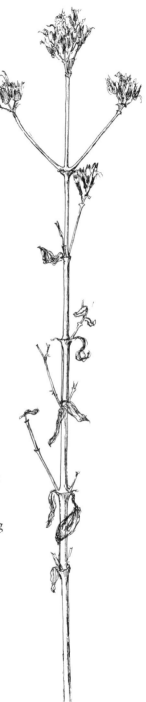

Bouncing Bet
Saponaria officinalis

(Soapwort)

Bouncing Bet does not last the winter at all well.
Its fruits generally look like a bedraggled mess, but
if you look closely, you might find an elongated
four-toothed capsule almost completely covered by
a ragged calyx. The best way to recognize Bouncing
Bet is by its stout, square stem, its stiff branches,
and the large bulges at its nodes.

Bouncing Bet grows in the sites which even
other roadside plants cannot tolerate. It is commonly found along railroad tracks, growing about
one to two feet tall. A perennial, it spreads by
underground runners and can form large colonies.
Its English and Latin names both come from the
fact that the juice inside the live plant forms a
lather when mixed with water.

BUTTERCUP FAMILY
RANUNCULACEAE

For fruits, most members of the Buttercup family have pods or achenes; a few have berries. The pods and achenes almost always have a long pointed tip, or beak; this is the best way to recognize the family. Many of them also have frizzy hairs which help in dispersion.

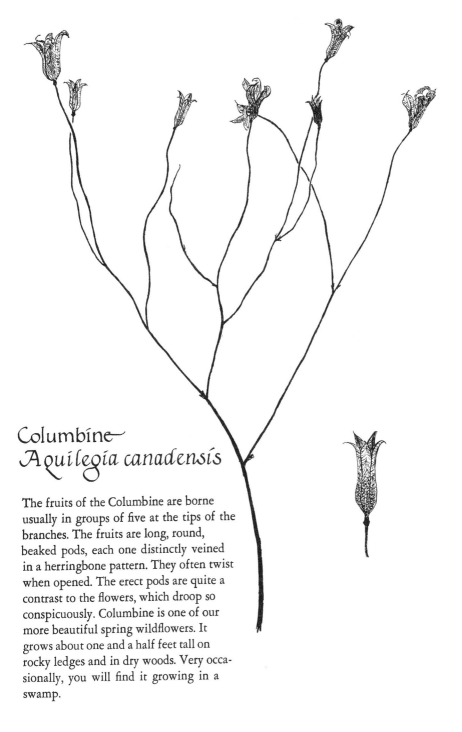

Columbine
Aquilegia canadensis

The fruits of the Columbine are borne usually in groups of five at the tips of the branches. The fruits are long, round, beaked pods, each one distinctly veined in a herringbone pattern. They often twist when opened. The erect pods are quite a contrast to the flowers, which droop so conspicuously. Columbine is one of our more beautiful spring wildflowers. It grows about one and a half feet tall on rocky ledges and in dry woods. Very occasionally, you will find it growing in a swamp.

one achene with hairs

Thimbleweed
Anemone virginiana
A. cylindrica

Thimbleweed grows one to four feet tall and has a long, slender stem which bulges at the nodes. The fruit is an achene with a little hook at the top, and each one has a spiral of fine fuzzy hairs attached to it. They are attached to an oblong receptacle, which remains after the fruits have blown away, covered with close wooly hairs. The stem is fuzzy towards the top and at the nodes, less so towards the bottom. In summer, Thimbleweed has toothed compound leaves which you might find in winter still drooping from the stem. The two species of Thimbleweed, *Anemone virginiana* and *A. cylindrica,* are not easy to tell apart.

Thimbleweed grows as a perennial in dry, open woods and sometimes on prairies.

receptacle covered with achenes

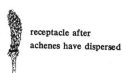
receptacle after achenes have dispersed

Tall Meadow~Rue
Thalictrum polygamum

The shape of this plant is not
very distinctive, but it might
attract your attention by its
height and by its smooth, shiny,
tan stem. The plant grows up to
seven or eight feet tall. In re-
lation to the rest of the plant,
the fruit is tiny; it is a beaked
achene less than a quarter of an
inch long, ribbed on the surface.
The fruits fall off very easily. Tall
Meadow-Rue grows in meadows
and swamps.

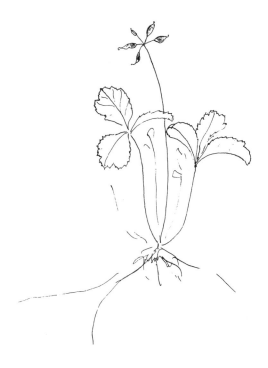

Goldthread
Coptis groenlandica

This is a small plant less than six inches high, that grows mainly in the mountains or in northern woods. You can identify it beyond a shadow of doubt by pulling one up and seeing its gold-colored rootstock. You can also identify it by its shiny evergreen leaves. The beaked fruits are borne in an umbel and are formed by late spring. Goldthread is a perennial. It spreads vegetatively, so you usually find the plant growing in patches.

Virgin's Bower
Clematis virginiana

This vine is easy to recognize by the fluffy hairs attached to its fruits. As is often characteristic of the Buttercup family, the fruit is a small, beaked achene, the beak in this case being prolonged into a trailing plume. Each flower produces many fruits arranged in a spiral on the small, button-like receptacle. If you find the plant early in the fall, you will probably find many perfect spirals, but the fruits quickly become detached from the receptacle, and later in the winter you will probably only find a mass of fluffy hairs twining among other bushes.

Virgin's Bower usually grows in areas with moist soil. Several species, varieties, and hybrids of the genus *Clematis* are grown for ornament.

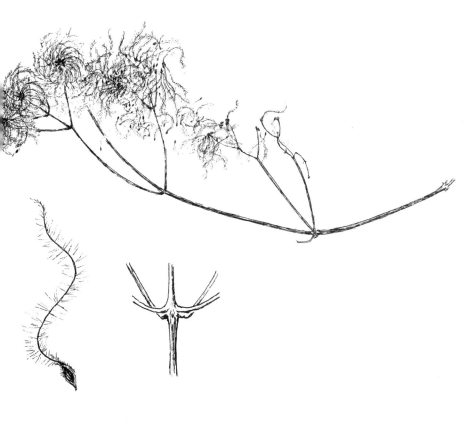

MUSTARD FAMILY
CRUCIFERAE

The Mustard family is characterized by a unique type of fruit known as a silique. *A silique is a two-parted fruit separated in the middle by a thin membrane. The seeds are found on both sides of the membrane. Siliques can take all kinds of shapes, as you will see in the following pages. Often, when the seeds ripen the outer walls of the silique fall off and the seeds are dispersed. Then all that is left on the plant is the membrane, which is often silvery or translucent. The "silver dollar" plants (p. 92), which people grow in their gardens for decoration are members of the Mustard family and the "dollar" is the membrane. As winter wears on, often the membrane too is destroyed, and all that is left is its skeleton.* Alliaria, *or Garlic Mustard (p. 89), is one plant that is often found this way. Mustard fruits are quite distinctive, and this is one of the few families in which the members are identified more easily by their fruits than by their flowers. Conveniently, they bear their fruits early in the growing season.*

When green, most mustards have a sharp, pungent taste which has made them valuable in agriculture. The condiment mustard is made from the oil in the seeds of the genus Brassica.

One species, Brassica oleracea, *has been in cultivation since the times of the Greeks and the Romans and has been bred to many varieties which include cabbage, kale, brussels sprouts, cauliflower, broccoli, kohlrabi, rape, rutabaga, and Chinese cabbage. Another species,* Brassica Rapa, *is the garden turnip. Watercress is a mustard,* Nasturtium officinale, *which was originally introduced from Europe for cultivation but now grows easily by itself in cool, fresh water. The garden radish (* Raphanus sativus *) is another member of the Mustard family, as is horseradish (* Armoracia rusticana *). In addition, many common garden flowers—stock, candytuft, alyssum, and woodruff— are members of the Mustard family.*

Shepherd's ~ Purse
Capsella Bursa ~ pastoris

The fruit of this plant is a heart-shaped silique, apparently shaped like a medieval purse. *Bursa-pastoris* means "purse of a shepherd," and *Capsella* is the diminutive of *capsa,* "box." Shepherd's-Purse is one of the earliest-blooming flowers in spring; it starts blooming in March in some areas. It forms its fruit almost immediately after blooming, so many of the spring-blooming plants you find may be quite beaten up by winter. However, the plant continues to bloom and set fruit through the summer and fall and even through the winter, if the weather is mild. Shepherd's-Purse grows as high as two feet, in gardens, waste places, and lawns.

Cow Cress
Lepidium campestre
(Field Cress, Field Peppergrass)

Pepper Grass
Lepidium virginicum
(Poor Man's Pepper)

The above collection of names should illustrate more than anything the unreliability of common names. These two species are similar in general size and appearance, but the fruits of *Lepidium virginicum* are flat and round, while those of *L. campestre* are deeper and more oblong. Both species can be found growing either as a single stalk or as practically a small bush. They last well when picked and make good winter decorations. They grow up to three feet high, sometimes as annuals and sometimes biennials. They are among the most common weeds of roadsides, fields, and parking lots.

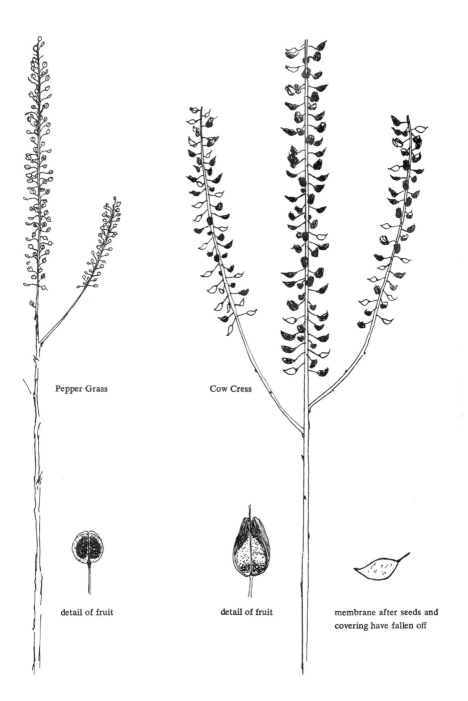

Pepper Grass

Cow Cress

detail of fruit

detail of fruit

membrane after seeds and
covering have fallen off

Mustard Family / Cruciferae **87**

whole fruit

fruit cut in half along membrane

membrane with seeds attached

membrane with no seeds

Penny Cress
Thlaspi arvense

Penny Cress is easy to recognize by the flat, papery wings which taper from the top to the bottom of the fruit and are notched at the tip. As with other mustards, you might find little left of the plant except for the flat pointed silvery membranes. Penny Cress grows anywhere from four inches to three feet high, sometimes as a single stalk, sometimes with many branches. It is found as an annual in fields and waste places, and is sometimes a troublesome weed (the Latin *arvense* means "of the fields").

Garlic Mustard
Alliaria officinalis

Not a very graceful plant, but quite common
in certain areas. The fruit is a long thin pod
that falls apart easily. By winter, not only have
the outer walls fallen off, but the membrane
that separated them has too, and all that is
left is the broken skeleton. The pods grow
one to two inches long, and the whole plant
grows up to four feet tall, with few branches.
You usually find it growing in patches in
open woods, along roadsides, or near buildings. The generic name is from the Latin
Allium, "onion," because the leaves have a
strong onion-garlic smell. *Alliaria* is a
biennial.

Yellow Rocket
Barbarea vulgaris
(Winter Cress)

The pods of Yellow Rocket are long and thin, with a pronounced beak at the tip. The plant flowers and sets fruit by early summer. The fruits are not very sturdy, so in winter you will mainly find the silvery membranes that are left after the fruit has split apart. They too are long and thin with the beak at the tip. The fruits are about one inch long, and the whole plant grows as high as three feet. It is usually quite bushy and has many fruits. Yellow Rocket is a common biennial growing in fields and along roadsides. The plant is named after St. Barbara. A different species of *Barbarea* was grown for food, and the seed was sown on St. Barbara's Day, in early December.

You might confuse *Barbarea* with *Alliaria* (p. 89), which also has long, thin pods, but the pods of *Barbarea* are smaller, more numerous, and more closely spaced together. *Alliaria* grows more often as a single stalk, and is not bushy like *Barbarea.*

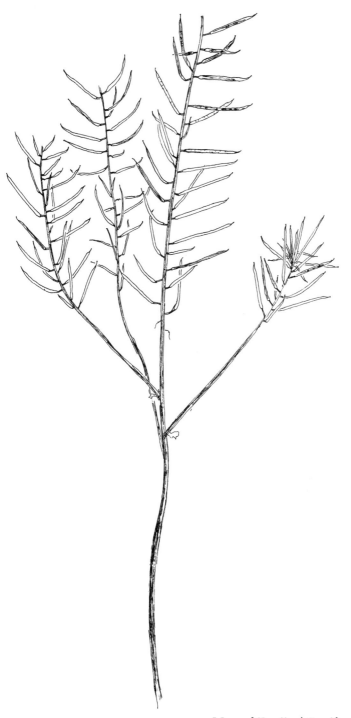

Silver ~ Dollars
Lunaria annua
(Money Plant, Honesty)

This plant does not really belong in this book, as it is a garden plant, but it sometimes escapes and grows on its own. It also re-seeds itself easily, so that you might find it in abandoned gardens. The translucent "silver-dollar," about the size of a quarter, is the membrane that separates the two halves of the fruit, but the rest of the fruit is just about as flat as the membrane, and the outer covering is brown and papery. The seeds are attached to the inside of this outer covering. The Latin name is from *luna,* "moon." Honesty grows up to three feet tall, as an annual or a biennial.

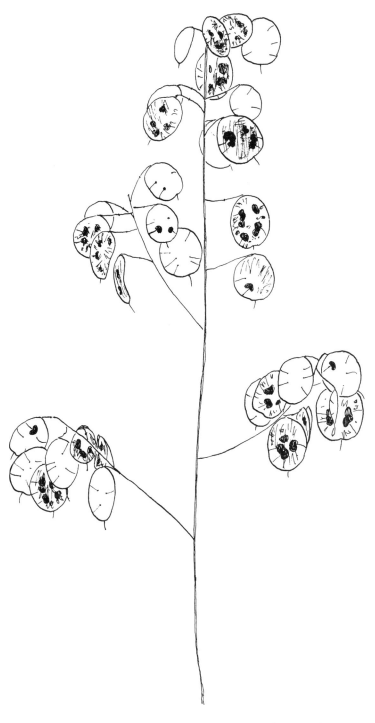

ROSE FAMILY
ROSACEAE

Although the characteristics of the Rose family are easy to spot in the summer, there are no constant characteristics that will help you to identify it in the winter. The fruits in the Rose family are quite varied. Many of them are fleshy—apples, peaches, and rose hips, and these will not concern us here. Several other species have achenes for fruits, and these achenes are often borne on a conspicuous conical or oblong receptacle. The achenes fall off early in the winter, leaving the receptacle, which you might find. Other species have for fruit a dry woody pod, usually fairly small. Spiraea, on p. 98, is one of these. The flower parts in the Rose family are always in fives; you might find five dried sepals under the fruit. Many members of the Rose family have compound leaves which might dry and remain on the plant, and they all have stipules—small leaflets at the base of the leaves. The stipules might also remain through the winter. Unfortunately, however, none of these characteristics are ones that you are sure to find, and none of them are exclusive to the Rose family.

The Rose family is important in daily life. It includes apples, peaches, pears, apricots, cherries, plums, and almonds, not to mention, of course, garden roses.

Rough ~ fruited Cinquefoil
Potentilla recta

From a distance, you would think that this perennial had round capsules, but up close, you see that nothing is left but the hairy bracts. The fruit is a head of achenes which are dispersed early in the season. The flower stalks branch oppositely, with a flower in each node.

Unfold some of the dried leaves on the main stem. You will see that they keep their five-fingered shape, which gives the plant its common name, Cinquefoil, French for "five leaves."

Cinquefoil grows one to two and a half feet high in old fields and along roadsides.

dried bracts and sepals

Tall Cinquefoil
P. arguta

Tall Cinquefoil has the same kind of opposite branching as Rough-fruited Cinquefoil (preceding page), and basically the same kind of flower remains. However, the dried flower heads of Tall Cinquefoil are bigger (about three-eighths of an inch across), fewer, and less rounded than those of Rough-fruited. Also, the flower branches stay close to the main stem, and the bracts are not hairy. In flower, this plant is distinguished by being the only white *Potentilla* of thirty-three northeastern species. It is a perennial which grows one to three and a half feet tall in dry woods, rocky soil, and prairies.

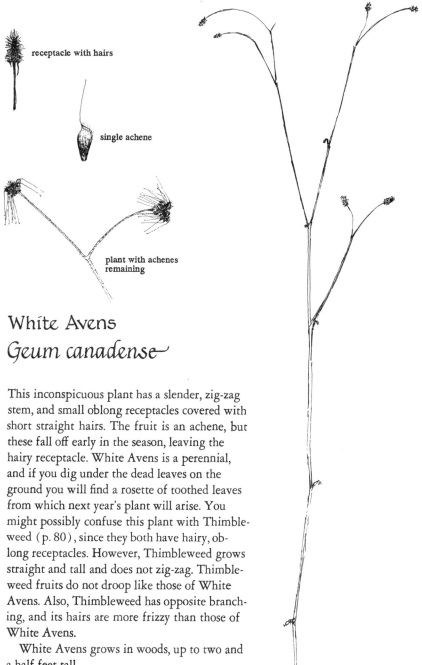

receptacle with hairs

single achene

plant with achenes
remaining

White Avens
Geum canadense

This inconspicuous plant has a slender, zig-zag stem, and small oblong receptacles covered with short straight hairs. The fruit is an achene, but these fall off early in the season, leaving the hairy receptacle. White Avens is a perennial, and if you dig under the dead leaves on the ground you will find a rosette of toothed leaves from which next year's plant will arise. You might possibly confuse this plant with Thimbleweed (p. 80), since they both have hairy, oblong receptacles. However, Thimbleweed grows straight and tall and does not zig-zag. Thimbleweed fruits do not droop like those of White Avens. Also, Thimbleweed has opposite branching, and its hairs are more frizzy than those of White Avens.

White Avens grows in woods, up to two and a half feet tall.

Hardhack
(Steeple-bush)

Meadowsweet

Spiraea spp.

Technically, *Spiraea* should not be included in this book, for it is a
shrub and not a herbaceous plant. Scratch the bark (when the plant is
freshly picked) and you will see that the stem is green underneath. This
means that the above-ground parts of the plant are still alive. The buds
along the stem will produce new leaves and shoots in the spring, unlike
the other plants in this book, which are dead from the ground up. However, because of its size and general growth form, *Spiraea* is often considered a "wildflower" rather than a shrub. It is a plant that you are likely
to notice and pick out in the field; therefore it is included in this book.

The two most common species are *Spiraea tomentosa,* Hardhack, and
S. latifolia, Meadowsweet. The fruit of each is a *tiny* pod, opening along
one seam, arranged in groups of five. Despite the similarity of the fruits,
the two species are easy to tell apart. First, all parts of *S. tomentosa* are
fuzzy, as the Latin name implies, while *S. latifolia* is completely smooth.
Second, the branches of *S. tomentosa* are shorter and more closely
spaced than those of *S. latifolia,* so that the whole plant looks more
narrow.

If you find a plant with the branching of *S. latifolia,* and with a little
bit of fuzz on the stem but none on the fruits, you have a third species—
Spiraea alba, also called Meadowsweet. Farther south, you might find
Spiraea corymbosa, which has the same tiny fruits in a flat-topped
cluster.

These various species of *Spiraea* grow from four to six feet tall in old
fields, swamps, and meadows throughout the Northeast. You might find
Spiraea growing as a single stalk or as a bush several feet in diameter.

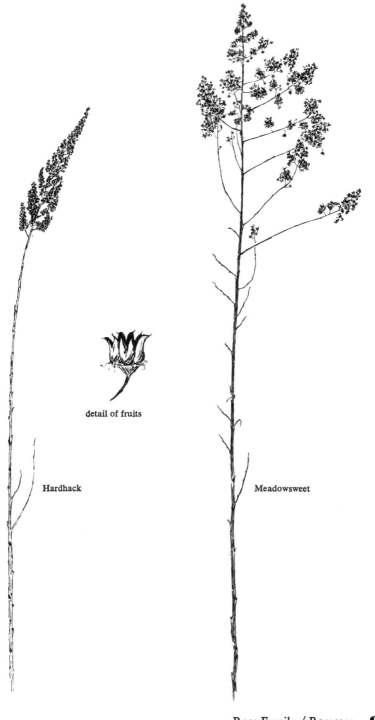

detail of fruits

Hardhack

Meadowsweet

Rose Family / Rosaceae **99**

 whole bur

 cross section of bur showing receptacle with achenes attached

 one achene

Avens
Geum laciniatum

This plant has no common name to call its own. There are many species of *Geum* which are all called Avens, so we are better off referring to this one by its Latin name. From a quick look at the fruit, you might confuse *Geum laciniatum* with Burdock (p. 184). However, *Geum* is much more delicate and usually smaller. *Geum* grows three and a half feet or less, while Burdock can grow as high as five feet. *Geum* branches at every node while Burdock branches emanate from one central stalk. Also, if you look closely at the two fruits, you will see further differences. Burdock burs stick to you much more fiercely than those of *Geum,* because each hook is part of a rough covering that surrounds the whole fruit. If you pull one hook from a Burdock bur, all the others come with it. Each hook of a *Geum* bur, however, is attached to only one fruit (an achene), and not to the other hooks. If you pull off one hook, the bur begins to fall apart.

The stems of *Geum* are slightly hairy throughout, but most obviously so at the bulging nodes. Each fruit also has a few hairs towards its tip. *Geum laciniatum* grows in wet places, meadows, and clearings, and along roadsides.

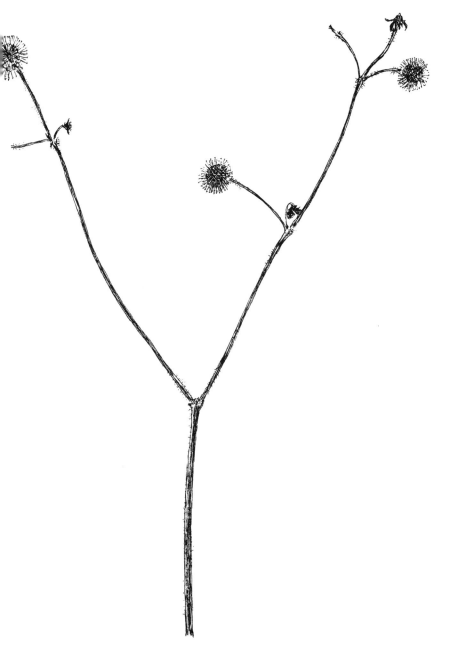

Agrimony
Agrimonia spp.

The Agrimony fruiting structures look like little tops, with several rows of hooked bristles around the middle. The upper section of the top has several grooves in it. This bristly structure is not the fruit itself but the dried calyx. The sepals have this same basic shape when the plant is in flower, except that the grooved section is smaller and expands as the fruit ripens. The pointed "tip" of the top consists of five sepals that were opened under the flower, but join together and close during the formation of the fruit. The fruit is an achene inside this structure.

There are several species of Agrimony, all quite similar. Some of them grow as high as six feet, but you most often see the plant at a height of three feet or so. It prefers thickets and open woods.

BEAN FAMILY
LEGUMINOSAE

Everyone knows what a pea pod looks like. A pea pod is a legume *as is a string bean, and a legume is the characteristic fruit of the* Leguminosae. *(The word legume is also loosely used to refer to any plant in the Bean family). Unfortunately, although the similarity between a pea pod and a string bean is instinctively obvious, the word legume is not so easy to define.* Gray's Manual *tells us this: "Legume. The fruit of the* Leguminosae, *{a circular definition} bilaterally symmetrical and produced from a unilocular ovary, one to many seeded, variously dehiscent or indehiscent. . . ."* Britton and Brown *tells us that a legume is "a dry dehiscent fruit derived from a simple ovary and usually dehiscing along two sutures." Neither of these definitions is much help to us. First of all, they contradict each other, for* Britton and Brown *says that legumes always open (dehiscent), while* Gray's *says that some do and some do not. The main distinguishing feature of a legume, in these two definitions, seems to be that it is derived from a one-celled (unilocular) ovary. This is fine, but by the time you and I see the plant, in winter, the ovary has turned into the legume, and we have no way of knowing how many sections it had. Furthermore, even if we had the live plant in summer, we would need a fine knife and a magnifying glass, if not a microscope, to determine the number of sections in the ovary. We can say this much, however, about dried legumes:*

- *They are bilaterally symmetrical.*
- *They open along one or two seams, the full length of the fruit.*
- *Many of them have seeds lined up along one edge, like a pea pod.*
- *Most of them are not partitioned on the inside.*

You will have to rely partly on your instinctive sense of the relationships between plants to know a legume when you see it, but this will be easier than you might think.

The commercial importance of the Bean family is well-known: peas, string beans, kidney beans, lentils, alfalfa, peanuts, and soybeans, the food of the future, are all legumes. Licorice and indigo dye are also made from legumes. Common garden plants include Albizzia *(the Silk-tree), and the wonderful-smelling vine* Wisteria. *Legumes are also important as soil enrichers, for they have the ability to "fix" atmospheric nitrogen into a form that plants can use.*

Bush · Clover
Lespedeza capitata

Bush-clover is easy to recognize by its fuzzy
brown heads of dried calyces. If you poke inside
the calyx, you might find a tiny hard pod. Bush-
clover is very common in fill areas, along railroad
tracks, and in other areas with sandy soil. It
grows two to five feet high.

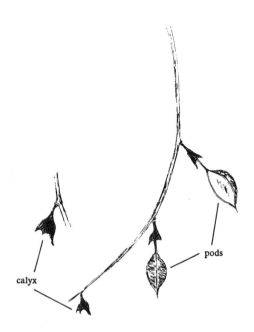

calyx

pods

Wild Indigo
Baptisia tinctoria

(Rattleweed)

This is a much branched, very scraggly perennial with shreddy bark.
Although the plant grows two to three feet high, you usually find it lying
down on the ground in dry, sandy places. The fruit is a small pod less
than a half inch long. The branches of the plant are very flexible, so that
when you shake it the pods knock against the other branches and the
plant rattles. Many of the pods fall off and leave only the calyx (see
detail). The names *Baptisia* and *tinctoria* come respectively from the
Greek and Latin words for "dye," for the plant was used in colonial
times to produce a blue dye. True indigo dye, however, comes not from
this plant but from an Asian plant, *Indigofera tinctoria,* also a member
of the Pea family.

Tick Trefoil
Desmodium spp.

(Tick Clover, Beggar's Ticks)

This is a mystery plant. You will probably more often find the little pods stuck to your clothes than you will ever see the whole plant. Although it grows up to six feet tall, the plant is inconspicuous, and it usually loses most of its fruits early in the season, as they stick to every passer-by. The pods attach themselves to you by means of tiny hairs with hooks too small to be seen with the naked eye.

The pods are made up of many sections, attached to each other in a chain (the Latin name comes from the Greek word, *desmos,* for chain). These sections separate easily, and you will probably find mainly single sections on your clothes. If you open one up, you will find a seed inside shaped like a lima bean.

There are many species of *Desmodium* which are not easy to distinguish. Most of them grow in woods, but some grow in meadows or thickets.

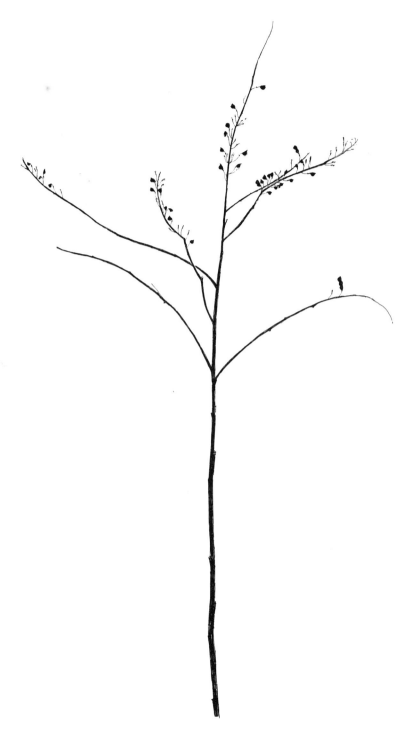

Wild Sensitive Plant
Cassia nictitans

Partridge Pea
C. fasciculata

(Golden Cassia, Prairie Senna)

At first glance, both of these beans look very much alike. They both have flat brown pods one to two inches long which twist into a spiral after they have opened. The stems and the pods of each are sometimes slightly hairy, sometimes smooth. Both species grow in dry, sandy soil.

The best way to distinguish the two species is by the length of the flower stalk. Partridge Pea (*C. fasciculata*) has an upcurving flower stalk about an inch long, while the pods of the Wild Sensitive Plant (*C. nictitans*) are borne on very short stalks close to the stem. Both species have *stipules*—little leaflets at the base of the flower stalk—but those of the Partridge Pea are larger and more obvious, even in winter. Partridge Pea is generally larger than Wild Sensitive Plant. It grows to a height of two and a half feet, while Wild Sensitive Plant rarely grows taller than one and a half feet; but you should not try to distinguish the two species merely on this basis. Partridge Pea has a wider distribution, growing as far north as Minnesota, Ontario, and Massachusetts, while Wild Sensitive Plant stays south of a line that goes through Missouri, Illinois, Indiana, and Ohio to Massachusetts. Both species are annuals.

Wild Sensitive Plant (*C. nictitans*) gets its name from the fact that its compound leaves fold up when you touch them. The Latin word *nictitans,* which means "winking," also applies to this habit.

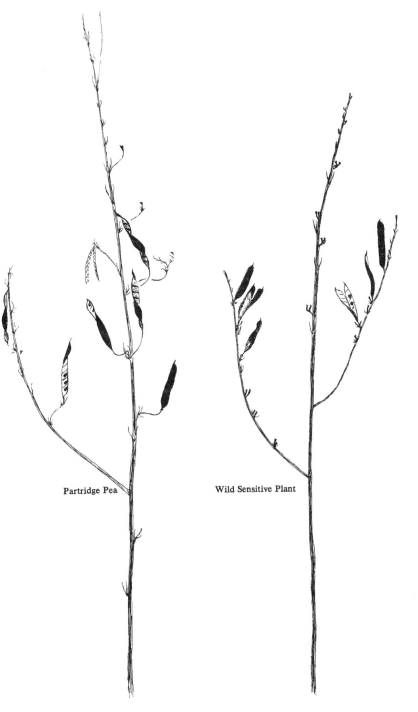

Partridge Pea

Wild Sensitive Plant

Bean Family / Leguminosae **111**

Everlasting Pea
Lathyrus latifolius

This plant is a close relative of the garden Sweet Pea, *Lathyrus odoratus*, and is also commonly grown in gardens. In flower, these two species are almost identical, except that our species has no smell, while the Sweet Pea is known for its fragrance. Everlasting Pea has escaped from gardens and has established itself frequently along roadsides and in vacant land. Although it is a vine, it does not need other plants for support, and you will often see it growing in a tangled heap in the middle of a roadside clearing. Sweet Pea sometimes escapes but rarely lasts.

The pods of the Everlasting Pea open into a tight corkscrew. They are smooth and tan and about three or four inches long. Notice the dried compound leaves, characteristic of the Bean family, and the winged stems, characteristic of the genus.

Lupine
Lupinus perennis

Lupine is most easily recognized by the very fuzzy
pods which twist into a spiral after they open. The
pods are clustered towards the top of the stem, which
is also somewhat hairy. The plant grows eight inches
to two feet tall in sandy soil, and although it is dis-
tributed throughout the Northeast, it is not common.
Other species of Lupine are very common in the
western states.

MALLOW FAMILY
MALVACEAE

Many plants in this family are hairy, and one that has been put to good use is Gossypium sp., *or cultivated cotton. The hairs that make cloth are found surrounding the seeds. Two other familiar members of the Mallow family are the garden shrub Rose-of-Sharon, and the garden Hollyhock. The fruit in the Mallow family is often a five-parted capsule, or it is made up of many separate sections arranged in a circle. Many plants in the family have sticky juice, such as Okra* (Abelmoschus esculentus) *and Marsh Mallow* (Althaea officinalis).

Rose Mallow
Hibiscus palustris

Rose Mallow has a very fuzzy stem and woody five-parted capsules that open like flowers. The insides of the capsules are lined with brown fuzz, and the sepals, if they are still left on the plant, are velvety to the touch. Rose Mallow grows three to six feet tall in various kinds of wet places. From Massachusetts south to North Carolina, you only find it in salty or brackish areas, usually where there is standing water. A Cattail marsh, or the shore of a seaside pond, are good places to look for it. From western New York to northeastern Illinois, however, the plant also grows inland in fresh-water marshes. Rose Mallow is a perennial with beautiful pink flowers that bloom late in summer.

If you have a Rose-of-Sharon bush (*Althaea rosea*) in your yard, you might notice that the fruits are very similar to those of Rose Mallow, as they are in the same family.

Velvet ~ leaf
Abutilon Theophrasti

(Piemarker)

This complex-looking fruit has many sections arranged around a central "spool." The individual sections fall off very easily. If you find the plant in late winter, or if it grows along a heavily travelled road, where the wind from the cars can knock it apart, you will probably find nothing left but the central spools. Each section is papery, with hairs along its bottom edge and a long curved beak on the outside. The stem and the sepals are velvety to the touch.

The growth form of Velvet-leaf varies considerably. It grows from two to five feet high, sometimes as a single stem, sometimes with many branches. It grows as an annual in cultivated fields, vacant lots, old fields, and other dry open places, and is more common the farther south you go.

PARSLEY FAMILY
UMBELLIFERAE

An umbel is like an inside-out umbrella. In an umbel, all the flower stalks radiate from one point at the tip of the stem. All members of the Umbelliferae bear their fruits in umbels. Many are in compound umbels, where each stalk of the umbel is topped with a second umbel. The fruit of the Umbelliferae is usually small and hard and looks like a seed. It is called a schizocarp, *which literally means "split fruit," because it splits in half when it is ripe. Often you will find half a fruit on the plant, or you might be misled into thinking that the fruits are borne in pairs.*

Members of the Umbelliferae have many distinctive characteristics. The fruits and the stem are often grooved and the stem is often hollow. The leaf bases completely encircle the stem and form a big bulge. In winter you can still see the bulge and the wraparound scar.

The Parsley family makes itself known to us in many ways. Carrots, celery, and parsnip are all biennial members of this family which you never see in flower, because you eat them in the first year of their life cycle. Many herbs are in the Parsley family—dill, parsley, caraway, chervil, coriander, lovage, and fennel. The family also provides us with poison. The hemlock which Socrates drank was not from a hemlock tree, but from an Umbellifer, Conium maculatum. Conium maculatum, *or Poison Hemlock, grows in our area as well as in Greece and can kill you very quickly. Thus, although so many of the members of this family are good to eat,* don't eat any wild Umbellifers until you are sure what they are.

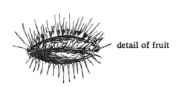
detail of fruit

Queen Anne's Lace
Daucus Carota

(Wild Carrot)

Queen Anne's Lace is a common plant of meadows, roadsides, and waste places. It usually grows to two or three feet in height. Its fruits are borne in an umbel that either radiates outward or sometimes curves back in on itself to form a little nest. The stem is grooved and slightly hairy, and the whole plant is somewhat coarse. If your plant still has fruits, they are very distinctive. They are ribbed, and each rib is lined with a row of bristles. The leaves of Queen Anne's Lace are very finely divided, and you might find some dried ones still on the plant.

Daucus carota is the same species as the garden carrot. The plant is a biennial, and in winter you can find rosettes of lacy, fuzzy leaves that will send up Queen Anne's Lace flowers the next summer. Smell the leaves to check their identity; if they smell like a mixture of carrots, parsley, and parsnips, they belong to *Daucus*. (They will probably not belong to either of the poisonous Umbellifers, Poison or Water Hemlock, for these both usually grow in wet places.) Dig up the rosette, and you will find a yellowy-white carrot.

Water Parsnip
Sium suave

Water Parsnip looks very much like Queen Anne's
Lace, but it grows in a completely different habitat—wet meadows, marshes and thickets. It also
grows taller—up to six feet—and its flower stalks
are completely straight, not curved in, like those of
Queen Anne's Lace. The most definitive way to tell
the two species apart is by the fruits, if any are left.
The Water Parsnip fruit is round or oval, and
ribbed, but when it splits in two, each half is
kidney-shaped. The Queen Anne's Lace fruit is also
ribbed, but each rib is lined with bristles. Water
Parsnip is a perennial.

(halved fruit

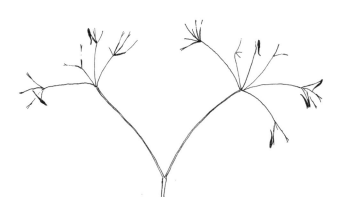

Sweet Cicely
Osmorhiza Claytoni

This is a spindly plant, and you will probably find a lot of plants broken down on the ground. You might not notice the plant except for its long hooked fruits which stick to your clothes. If you look at these fruits very closely you will see that they are ribbed like the fruits of other members of the Parsley family. They have up-curved hairs along the ribs which help them stick to you.

You might find Sweet Cicely with no fruits on it, but you can still recognize it by its umbels and by its spindly aspect. Also, if you look at the extreme tips of the flowering stalks, you will see that many of them are split in two at the top.

The Latin name *Osmorhiza* comes from the Greek *osme,* "scent," and *rhiza,* "root," for the roots and sometimes the leaves smell strongly of anise. Commercial anise, however, comes from another Umbellifer, *Pimpinella anisum.* Commercial licorice, which smells similar, comes from a member of the Bean family, *Glycyrrhiza glabra* (glycys is Greek for "sweet").

Sweet Cicely grows in moist, shady places, where it usually forms large patches. It grows one to two feet high.

Cow Parsnip
Heracleum maximum

There is no mistaking Cow Parsnip. It grows as tall as nine feet and
has all the textbook characteristics of a member of the Parsley family.
The small, flat, heart-shaped fruits are borne in large compound umbels,
and you can see bulges at each node, caused by the wraparound leaves.
The stem is stout, hollow, and deeply grooved. If you want to collect
Cow Parsnips while the fruits are still on them, be sure to go out early
in the fall or winter. The plant grows in damp soil; low meadows and
river edges are good places to look for it.

The specific Latin name *maximum* is an obvious reference to the
size of the plant. The English name Cow Parsnip probably came about
because of the dung-like smell of the flowers.

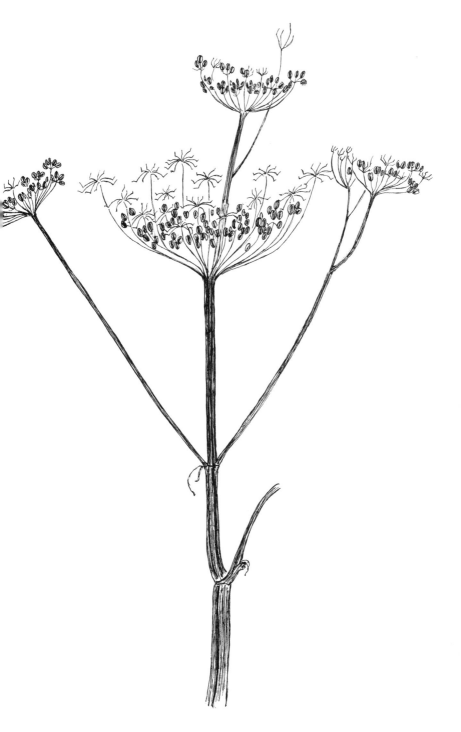

Parsley Family / Umbelliferae **123**

Meadow Parsnip
Pastinaca sativa

This is the same species as the garden parsnip that we eat. Like many members of the Parsley family, Meadow Parsnip is a biennial, and the root that we eat is formed during the first year's growth. The tall flowering shoot is produced in the second year. Meadow Parsnip usually grows in patches. Find a patch of the dried flowering shoots and then look on the ground for rosettes of deeply lobed, almost compound leaves. If the leaves are very finely divided (i.e., lacy) and devoid of hairs, DON'T GO NEAR THEM, because they might be Poison Hemlock, which can kill you in a very short time. But if you are confident that you have found the right rosettes, dig them up and you will find parsnips.

There exists some confusion as to whether Meadow Parsnip is poisonous, even though it is a variety of the species which includes the cultivated parsnip. Older books will tell you that Meadow Parsnip, or even garden parsnip gone wild, is poisonous, but more recent books say that this belief is nonsense. The leaves, however, can produce a fairly severe rash in some people, so you are probably best off leaving this plant alone.

The flowering shoot of the Meadow Parsnip grows quite tall—up to six feet—and usually has many branches. You can find it in fields, along roadsides, and in waste places. It has all of the characteristics of a member of the Parsley family—fruits borne in compound umbels, grooved stems, and wraparound leaf scars. You might have a hard time distinguishing Meadow Parsnip from Cow Parsnip (page 122), for they both grow tall, with similar inflorescences and smooth, deeply grooved stems. Their fruits are somewhat different, but they vary considerably, so this is not a reliable characteristic. They do, however, have a slightly different overall appearance. Meadow Parsnip tends to be more scraggly than Cow Parsnip as the winter wears on, and it is usually more diffusely branched. You can also distinguish the two most easily by their habitat. As the name implies, Meadow Parsnip grows in meadows and fields, and along roadsides, while Cow Parsnip is usually found in wet places. In flower, the two are very easy to tell apart, because Meadow Parsnip has yellow flowers while those of Cow Parsnip are white and smell like cow dung.

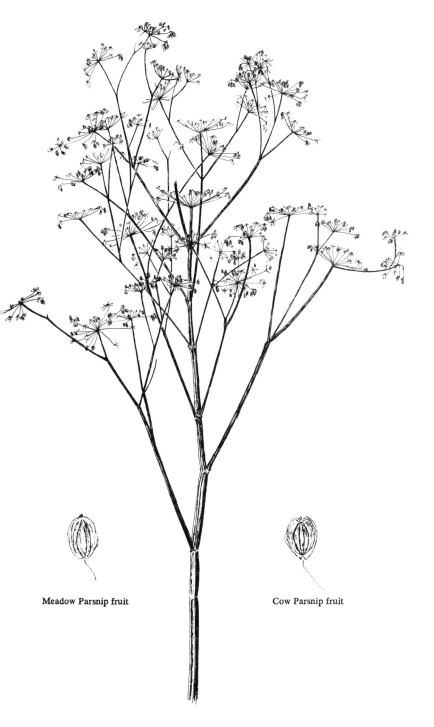

Meadow Parsnip fruit

Cow Parsnip fruit

WINTERGREEN FAMILY
PYROLACEAE

Most of the genera in this family are small plants that grow in oak or evergreen woods. They like shade and acid soil, conditions which are discouraging to most species. Some of them survive these conditions because they are saprophytes or parasites, which obtain their nutrients from other dead or living organisms.

All the members of the Wintergreen family have a four- or five-parted woody capsule. They are all perennials, and many of them have evergreen leaves.

Indian Pipe
Monotropa uniflora

Indian Pipe is a small perennial plant, usually about six inches high, that grows along the forest floor. You often find it in a dark pine or hemlock forest where few other species survive. The reason the plant can tolerate this dense shade is that it does not use the energy from sunlight to make its food the way other plants do. Rather, it is a parasite, extracting its food from fungi in the humus. Leaves, therefore, are an unnecessary part, since the main function of leaves is to catch sunlight to make food. Indian Pipe has only little scales on the stem, which persist through the winter.

The fruit of the Indian Pipe is a four- or five-parted woody capsule about one-half to one inch across. Although the flowers nod towards the ground, the capsule is erect.

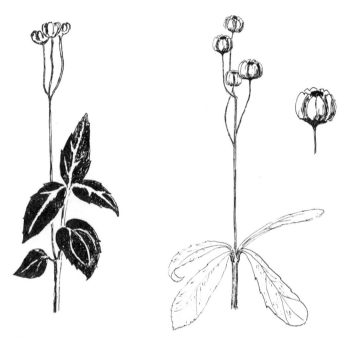

Pipsissewa
Chimaphila umbellata

Spotted Wintergreen
C. maculata

You will always find whorls of toothed evergreen leaves still attached to these small perennials. Both grow less than one foot high and crawl along the forest floor with spreading underground rootstocks. Rising from the evergreen leaves is a flower stalk with a few small, erect four- or five-parted capsules. Although the fruits are similar, the two species are easy to tell apart by their leaves. Those of Pipsissewa are a solid green and are relatively long and thin compared to those of Spotted Wintergreen. The Spotted Wintergreen leaves are easily recognized by their white veins. You will find both species of *Chimaphila* growing in dry woods where oak, hickory, and pine predominate.

Shinleaf
Pyrola elliptica

Shinleaf is a small plant, never more than a foot high, that grows mainly in unusual habitats—in the mountains and in bogs. You will find it rarely in woods elsewhere. Most of the species of this genus are northern, growing up into the Yukon. None of the species grow any farther south than West Virginia.

Shinleaf has rounded basal evergreen leaves and five-parted capsules that droop from the stem in a raceme. One way to recognize the flowers in summer is by their long curving pistils. These sometimes dry and last the winter. The Shinleaf seeds are minute, like dust.

MILKWEED FAMILY
ASCLEPIADACEAE

The Milkweed family has two constant winter characteristics:

1) The fruit is a pod. This means that it is bilaterally symmetrical and opens its full length along one or two seams.

2) The seeds have tufts of silky hairs attached to them.

When they are alive, plants in the Milkweed family have a milky juice.

Common Milkweed
Asclepias syriaca

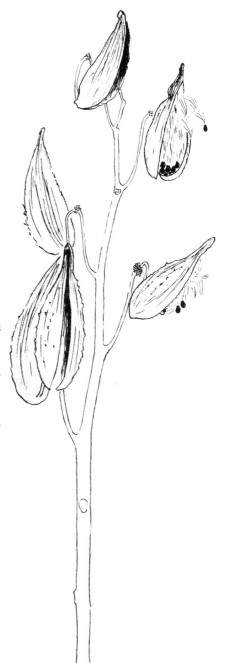

These warty pods are easy to recognize. They are fat at the bottom, pointed at the top, and open down the middle. You might find some pods with the flat, brown, silky-tufted seeds still packed inside them. The pods are often a shiny yellow on the inside. If you find pods shaped like these, but without warts, you have found a very similar species, *Asclepias purpurascens,* or Purple Milkweed. Milkweed is very common in old fields and waste places, and you have surely picked it in summer and watched the sticky milky juice come out. The plant grows three to five feet high.

The young leaves, the flower buds, and the young pods are all good to eat, but they must be boiled, drained off, and boiled again in fresh water two or three times in order to get rid of the bitter juice.

Milkweed flowers are borne in an umbel, but for some reason only one flower from each umbel develops into a fruit. You can still see the knob of scars left by the flower stalks.

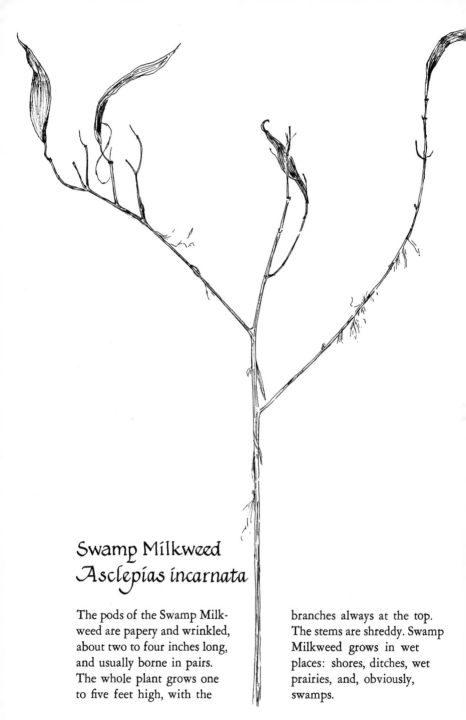

Swamp Milkweed
Asclepias incarnata

The pods of the Swamp Milkweed are papery and wrinkled,
about two to four inches long,
and usually borne in pairs.
The whole plant grows one
to five feet high, with the
branches always at the top.
The stems are shreddy. Swamp
Milkweed grows in wet
places: shores, ditches, wet
prairies, and, obviously,
swamps.

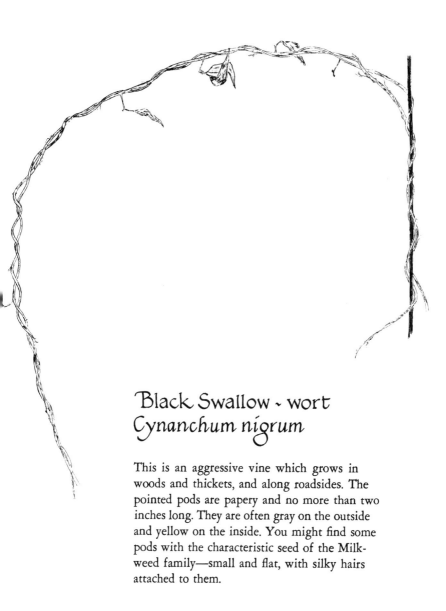

Black Swallow - wort
Cynanchum nigrum

This is an aggressive vine which grows in woods and thickets, and along roadsides. The pointed pods are papery and no more than two inches long. They are often gray on the outside and yellow on the inside. You might find some pods with the characteristic seed of the Milkweed family—small and flat, with silky hairs attached to them.

MORNING-GLORY FAMILY
CONVOLVULACEAE

Many of the plants in this family are vines, and many of them have smooth round capsules for fruits. One of the members of this family is the Sweet Potato, Ipomoea Batatas. *The Latin name of the family comes from the Latin* convolvere, *to roll together, in reference to the twining stems and to the habit of the flowers to twist when closed.*

Hedge Bindweed
Convolvulus sepium

(Wild Morning-Glory)

This is a vine with many similar relatives. It sometimes climbs onto objects or other plants and sometimes crawls along the ground, in vacant lots and along roadsides. Its fruit is a smooth round capsule about one-quarter to one-half inch wide nestled among dried bracts.

Dodder
Cuscuta spp.

(Love-vine, Strangle Weed)

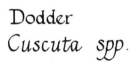

Dodder is a parasitic vine which you will always find wrapped around another plant. A Dodder seed on the ground germinates like the seed of any other plant, but the young shoot quickly attaches itself to a host plant by means of little suckers. Once the suckers are established, the Dodder roots die, and the plant loses all contact with the soil. It obtains all its nutrients through the suckers and is completely dependent on its host. The Dodder fruit, which is illustrated here, is a roundish capsule, less than a quarter of an inch wide. Occasionally, the seeds germinate right in the capsule, eliminating altogether the soil stage of the plant's life, and eliminating the time spent in locating a host plant.

There are many species of Dodder, some of which have a definite preference for certain host species. *Cuscuta Epilinum,* or Flax Dodder, grows only on cultivated flax and can damage it seriously. *Cuscuta Epithymum,* Clover Dodder, grows on plants in the Bean Family and can also be a serious pest. Most of the wild species which we see, however, grow on almost any kind of plant and in many kinds of habitats.

The fruit of the Dodder is similar to that of the Morning-glory (p. 135) but much smaller. Its stems are very slender, almost thread-like. In summer, you can recognize Dodder most easily by the intense orange color of its stems, which you see twining from one plant to another. The plant has no leaves and has very inconspicuous flowers.

MINT FAMILY
LABIATAE

The Mint family is one of the easiest to recognize. Every one of its members has a square stem and opposite branches. Many of them smell good in summer and winter. All the flowers have a fused calyx, often long and tubular, and usually toothed. After the flower petals dry and disappear, the calyx dries and remains through the winter. Inside the calyx are the fruits—four little nutlets—but these might be dispersed by the time you find the plant. Since the nutlets are hardly distinctive, the calyx is the feature you will use most often to tell one mint from another.

The Mint family includes many herbs: spearmint, peppermint, basil, rosemary, sage, thyme, oregano, lavender, hyssop, horehound, and catnip, as well as the bright red garden flower Salvia. *If you have any of these growing around your house, go take a look at them in winter or summer and you will see the family characteristics mentioned above.*

Selfheal
Prunella vulgaris

Selfheal is variable in height. Since it often grows on lawns or roadsides, where it is mowed, you might find it a few inches high; but left untended, it can get as high as two feet. Generally, however, it is a small plant. The calyces are in a spike at the tip of the stem and they are very hairy. Each calyx has two lips of unequal size, one almost fan-like. It is a perennial.

Blue Curls
Trichostema dichotomum

(Bastard Pennyroyal)

This is a small plant. It can grow up
to two and a half feet, but is usually
less than eight inches high. It grows
along railroad tracks, and in similar
places with poor, sandy soil. The plant
has many upcurving branches and a
papery, almost transparent calyx with
five teeth. The three lower teeth are
longer than the two upper. This plant
has a curious habit of somehow turning
its calyx upside down between summer
and winter, for when the plant is alive,
the three long teeth are on top. The
whole plant is covered with fine hairs.
Blue Curls is an annual.

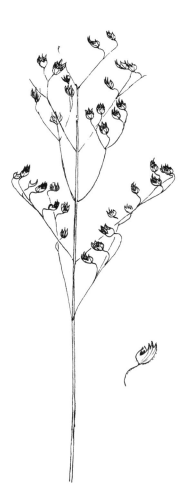

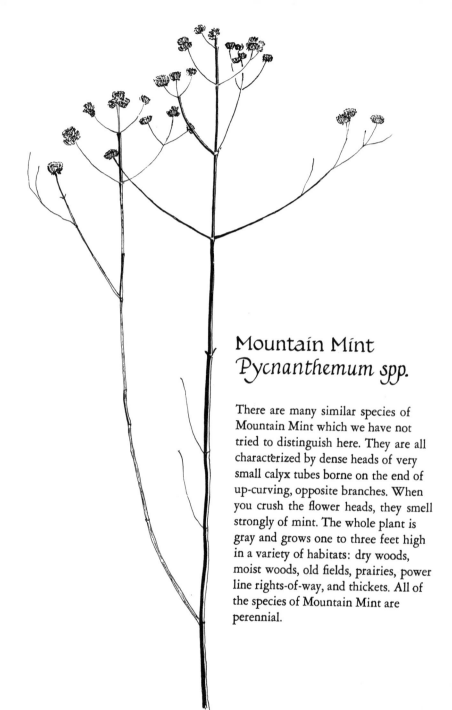

Mountain Mint
Pycnanthemum spp.

There are many similar species of
Mountain Mint which we have not
tried to distinguish here. They are all
characterized by dense heads of very
small calyx tubes borne on the end of
up-curving, opposite branches. When
you crush the flower heads, they smell
strongly of mint. The whole plant is
gray and grows one to three feet high
in a variety of habitats: dry woods,
moist woods, old fields, prairies, power
line rights-of-way, and thickets. All of
the species of Mountain Mint are
perennial.

detail of calyx

Wild Basil
Satureja vulgaris
(Savory)

Wild Basil is easy to recognize, because all parts of
it are so densely hairy. It is usually a small plant
which you will find growing in among other plants,
but it can grow as high as two feet. Like many other
members of the Mint Family, it has basal branches
that creep along the ground and send out roots, thus
insuring the spread of the plant. This species is perennial and you can often find the still-green runners
under the dead leaves. Because of this spreading habit,
you will usually find the dried flower stalks in patches.
Wild Basil is not really a basil; in fact, its leaves have
very little smell, but it is in the same genus as the
cultivated Summer Savory (*Satureja hortensis*).

The calyx tubes are arranged in dense head-like
clusters around the stem. Each tube has prominent
veins lined with hairs, and five pointed teeth.

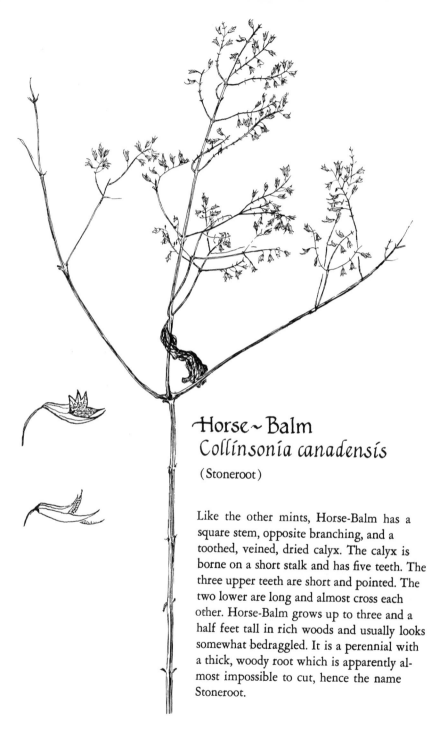

Horse~Balm
Collinsonia canadensis
(Stoneroot)

Like the other mints, Horse-Balm has a
square stem, opposite branching, and a
toothed, veined, dried calyx. The calyx is
borne on a short stalk and has five teeth. The
three upper teeth are short and pointed. The
two lower are long and almost cross each
other. Horse-Balm grows up to three and a
half feet tall in rich woods and usually looks
somewhat bedraggled. It is a perennial with
a thick, woody root which is apparently almost impossible to cut, hence the name
Stoneroot.

Water~ Horehound
Lycopus sp.
(Bugleweed)

Not a very interesting plant, but one commonly found in wet places. Like other mints, the plant has dried calyces, but they are so small that you can hardly distinguish them, and they look like little tufts encircling the stem. Water-Horehound grows six inches to two feet tall, as a single stem or with a few branches. It is a perennial.

Pennyroyal
Hedeoma pulegioides
(Mock Pennyroyal)

Although Pennyroyal is not found very frequently, I have
included it here because it is sure to attract your attention by its over-
powering mint smell, perhaps even before you see it. If you just brush
against it or step on it, you will release the smell, which will fill the air
around you. The plant is small—usually less than one foot—and grows
in very poor, dry soil. It likes a bed of crushed rock such as you would
find in an old gravel pit or on a new construction road.

The dried calyces are arranged in delicate whorls around the stem.
Each calyx has five teeth, the two lower ones longer and more narrow
than the three upper. The two lower teeth are slightly hairy, and the
throat of the calyx is almost closed off by short hairs. Pennyroyal
is an annual. If you have found the dried plant in winter and want to
see it in summer, you will probably find it not in the exact same spot as
the dried plants but close by.

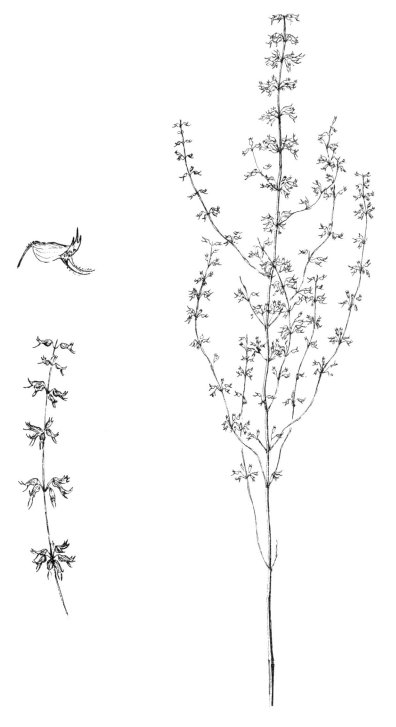

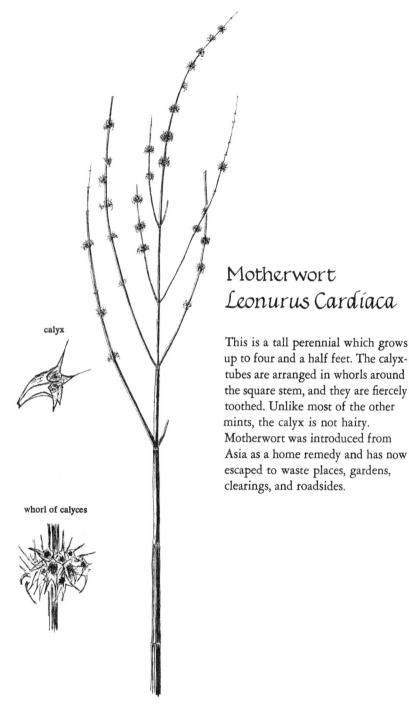

calyx

whorl of calyces

Motherwort
Leonurus Cardiaca

This is a tall perennial which grows
up to four and a half feet. The calyx-
tubes are arranged in whorls around
the square stem, and they are fiercely
toothed. Unlike most of the other
mints, the calyx is not hairy.
Motherwort was introduced from
Asia as a home remedy and has now
escaped to waste places, gardens,
clearings, and roadsides.

Bergamot
Monarda spp.

(Horsemint, Bee-Balm,
Oswego Tea)

If you crush these flower heads, they
have a wonderful smell, not sharp
and pepperminty like other mints.
but quite rich. Like other mints,
Bergamot has square stems and opposite branching, and all that remains after flowering is a dried
calyx tube. These tubes are crowded
into dense, rounded heads, usually
at the tip of the stalk. Bergamot
grows quite tall—up to four or five
feet. It is most common in northern
areas, although it can be found
throughout the Northeast. There are
several species not worth differenti-
ating here, and they have various
habitats: old fields, clearings, moist
woods, and thickets. Some species
are annuals and some are perennials.
The Bergamot flowers are quite
vivid, and the plant is often grown
in gardens.

head of calyx tubes

individual calyx

Mad~Dog Skullcap
Scutellaria lateriflora

The dried calyx of this perennial is somewhat different from that of the other mints .First, its edges are smooth, not toothed. Second, the calyx has two distinct halves which separate. The top half, with a distinctive little hump, falls off easily. This leaves the bottom half, which looks like a little scoop (the name comes from the Latin *scutella,* "dish"). The stem of the plant is square, with opposite branching, and the flowers are usually found in pairs along the stalk. Mad-Dog Skullcap grows about one to two feet tall, in wet soil.

The common name has two separate origins: "Skullcap" refers to the hump on the calyx, which apparently looks like a kind of cap which the Romans used to wear. "Mad-Dog" refers to the fact that the plant was once used as a remedy against rabies.

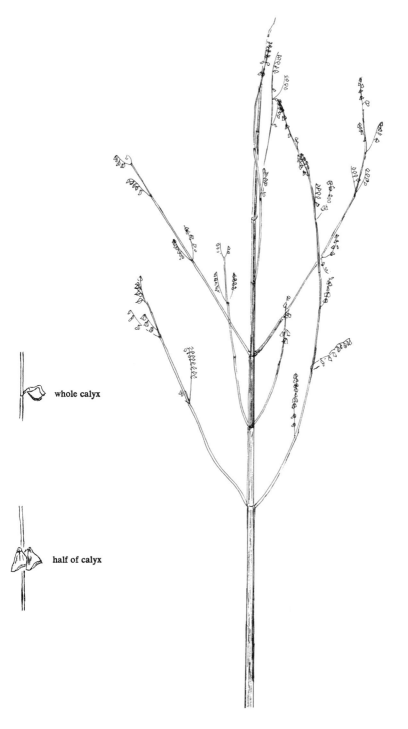

whole calyx

half of calyx

TOMATO FAMILY
SOLANACEAE

Structurally, the two representatives of the Tomato family in this book have little in common. The fruit of one is a yellow berry, and the other is a thorny capsule. Chemically, however, they have in common the presence of alkaloids, a class of chemicals which produce strong reactions in animals. The drug atropine is obtained from a member of the Tomato family, Atropa belladonna, *and other plants in the family are hallucinogenic. Many are deadly poisons. Tobacco* (Nicotiana Tabacum), *which you might consider a poison or a useful plant, is in the Tomato family, as are eggplants, potatoes, and, of course, tomatoes. Sunburned potatoes are known to be poisonous, and tomatoes were not eaten for a long time after their introduction to Europe, for they were thought to be poisonous. We laugh when we hear that now, but actually the Europeans were smart not to eat them, considering their botanical relatives.*

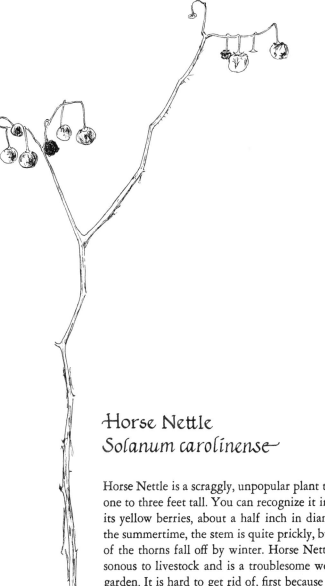

Horse Nettle
Solanum carolinense

Horse Nettle is a scraggly, unpopular plant that grows
one to three feet tall. You can recognize it instantly by
its yellow berries, about a half inch in diameter. In
the summertime, the stem is quite prickly, but many
of the thorns fall off by winter. Horse Nettle is poisonous to livestock and is a troublesome weed in the
garden. It is hard to get rid of, first because the thorns
make it hard to pull up by hand, second because it
is a perennial, so if you don't dig up the whole root,
it will come right back again next year. But a field full
of these berries in winter can be a beautiful sight.

   This plant is in the same genus as the garden
potato, *Solanum tuberosum.*

Jimson Weed
Datura Stramonium

(Thorn-Apple)

This sinister-looking annual is unmistakable because of its spiny four-parted fruits. Notice the skirt-like calyx and the lacy membrane that divides the sections of the fruit. *Datura* is frequently mentioned in a recent popular book, *The Teachings of Don Juan: A Yaqui Way of Knowledge,* by Carlos Castaneda. This book tells about a Yaqui Indian's use of a western species, *Datura inoxia,* as a hallucinogen. Reading this book has prompted some easterners to try to get high on *Datura Stramonium.* Don't. All parts of the plant are extremely poisonous. In *Poisonous Plants of the United States and Canada,* by John M. Kingsbury, we find this description of the symptoms of *Datura* poisoning: "Subjects become delirious, incoherent, and perform insensible motions, commonly picking at imaginary objects on themselves or in the air. Temperature may be elevated, and the heartbeat rapid and weak. Subjects may become violent and dangerous to themselves and others. If the poisoning progresses further, convulsions appear, followed by coma." The final result is often death.

Actually, all species of *Datura* are equally poisonous, due to the presence of alkaloids, and you should not experiment with any of them. The Indians have doubtless learned the proper dosage through long experience. In addition to the American Southwest, *Datura* is also used by many Indian tribes in South America, where it is reported that "the intoxication is marked by an initial violence so furious that the partaker must be restrained."

Datura is poisonous to livestock as well as to people, but animals do not like it and leave it alone if they can. This might explain why barnyards and pastures are among the most common sites for *Datura.* Livestock poisoning occurs only if there is no other forage available for the animals.

The English name of the plant is derived from an incident that took place in 1676 in Jamestown, Virginia. British soldiers were sent to quell a local rebellion and in the course of the battle ran out of food. They ate the Jimson Weed that was growing in the area and were all poisoned. "Jimson Weed" is a corruption of "Jamestown Weed."

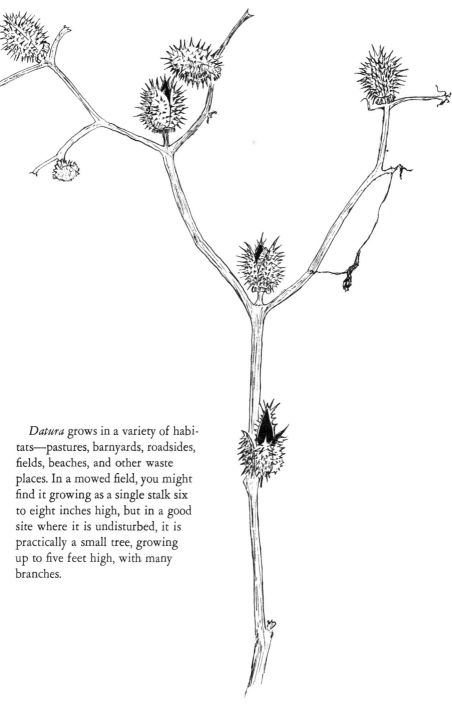

Datura grows in a variety of habitats—pastures, barnyards, roadsides, fields, beaches, and other waste places. In a mowed field, you might find it growing as a single stalk six to eight inches high, but in a good site where it is undisturbed, it is practically a small tree, growing up to five feet high, with many branches.

FIGWORT FAMILY
SCROPHULARIACEAE

The most constant characteristic of the Figwort family, or the "Scrophs," as they are often called, is the fruit—a two-parted capsule. Other characteristics of the "Scrophs" vary—some plants have square stems and some have round; some have opposite branching, some alternate. But this two-parted capsule, which often looks like a turtle's head, is characteristic of every "Scroph" in this book.

Common Mullein
Verbascum Thapsus

Mullein grows as high as six feet and is exceedingly
common along roadsides, on gravel banks, and in any
open' area where the soil is dry and rocky. The plant
was introduced from Europe but is spreading widely,
and you occasionally find it growing in woods too.
It is a biennial and during its first year, it produces a
rosette of large, flannelly, light-green leaves which
you can easily find in winter. In its second year it produces the tall flowering spike. Although the fruits are
very closely crowded together and are somewhat hard
to distinguish, if you look closely you will see the
two-parted capsule characteristic of the Figwort family. The stem is quite rough and fuzzy.

Moth Mullein
Verbascum Blattaria

Moth Mullein is easily recognized by its round, bead-like capsules lined up along a tall, straight stem. You might find the capsules opened or unopened, each one on a short upcurved stalk. The plant grows three to five feet high, sometimes branched, sometimes not. It is a biennial found in fields and vacant lots, and along roadsides.

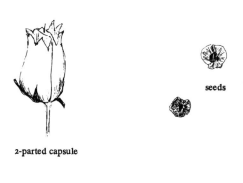

seeds

2-parted capsule

Butter ~ & ~ Eggs
Linaria vulgaris

This plant has two-parted capsules like the rest of the members of the Figwort family, but you might not recognize it instantly as a member of the Figwort family, since the capsules do not open turtle-head fashion like those of many other "Scrophs." Also, the Butter-and-Eggs capsules disintegrate easily, so they are not so easy to distinguish. The capsule is divided in half by a thin wall and is toothed at the top of each half. The alternate branches twist and curve around each other, and the whole plant looks a little unorganized.

Butter-and-Eggs is a very common perennial. It spreads by underground rhizomes and forms colonies in fields and waste places. Although it can grow as high as three feet, you rarely find it taller than a foot or so, perhaps because you see it most often on roadsides, where it gets periodically mowed, sprayed, and generally subjected to abuse.

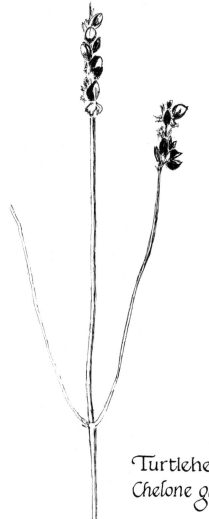

Turtlehead
Chelone glabra

These fruits do not hold up very well against the winter, but if you look closely you can find two-parted capsules that, surprisingly enough, look like a turtle's head. (It is more likely, however, that the name of the plant comes from the flowers, which also look like turtles' heads.)

The stem is square, and the fruits are clustered in a spike at the tip. The branching is opposite. Turtlehead grows in wet places as a perennial. It reaches a height of about three feet.

Common Speedwell
Veronica officinalis

Veronica is one of the few plants in this book that crawls along the ground and does not stand erect. You can recognize it easily by the fuzzy green leaves that stay on the plant through the winter and by the tiny, two-parted, heart-shaped capsules, which are almost flat. All parts of the plant are fuzzy. *Veronica officinalis* grows in woods, fields, and dry, rocky places. Although the main stem stays low to the ground, the flower stalk does climb upwards for a few inches, sometimes as high as a foot. *Veronica officinalis* is a perennial.

There are many species of *Veronica* in our area. Most of them are crawlers, like *V. officinalis,* but a few grow upright and become quite large. None are as common or as persistent as *V. officinalis,* but they all have similar heart-shaped fruits.

There are four species—*Veronica serpyllifolia, V. agrestis, arvenis,* and *V. persica*—which you might have growing in your lawn. They can turn a whole lawn blue in early spring. If you poke through the grass late in the spring, you might find some of their characteristic heart-shaped capsules.

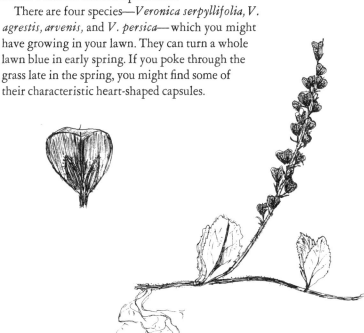

False Foxglove
Gerardia virginica

This is a plant of ambiguous characteristics. Its stem is almost square, but not quite; its branching is usually opposite, but not always. However, it is still fairly easy to recognize by its tall size (up to five feet), and its two-parted capsules with a beak at the tip of each section. Its habitat is also a clue, as the plant grows only in oak woods, where it is said to be parasitic on the roots of the trees. It is a perennial.

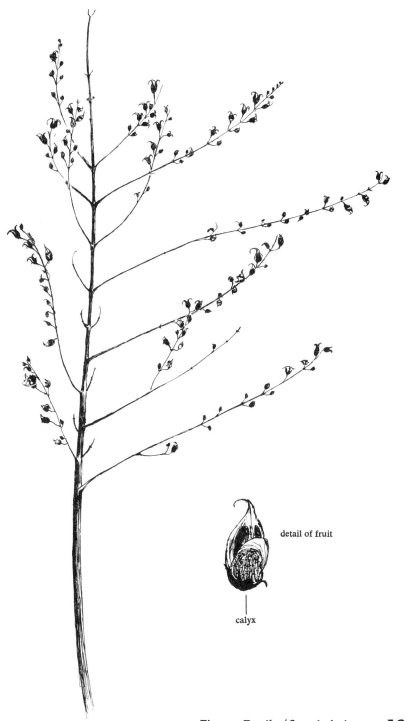

detail of fruit

calyx

Monkey Flower
Mimulus ringens

It can be hard to make out the structure of this fruit, because the prominently ribbed calyx often covers the capsule completely. The calyx is also easily torn up, so the whole fruiting structure is somewhat ragged. However, after a little bit of looking you can probably distinguish a somewhat flattened, two-parted capsule with two notches at the top. The walls of the capsule curve in towards the top. The plant usually has many fruits, borne on opposite, up-curving stalks, and it grows almost four feet tall. The stem is grooved on two sides, so it looks square although it really is not. Monkey Flower grows in wet places as a perennial. Both its English and Latin names refer to the flowers, which look like a little face.

capsules

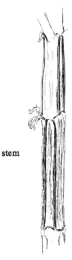

stem

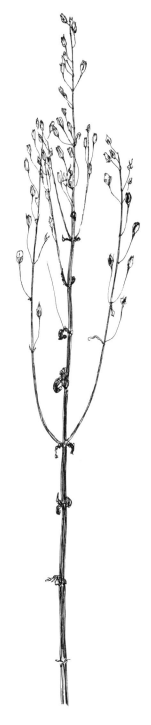

Figwort
Scrophularia spp.

These two perennial species are quite similar, and there is some doubt
in the botanical literature as to whether they can be safely told apart in
winter. They both grow quite tall—at least six feet—but *S. marilandica* sometimes reaches a height of nine feet. They both have square
stems, opposite branching, and small, two-parted fruits. According to
those who think they can be differentiated, the fruits of *S. marilandica*
are slightly smaller, and rounder than those of *S. lanceolata* (see illustration). Also, the stems of *S. marilandica* are grooved, while those of
S. lanceolata are not. *S. marilandica* is more of a woodland plant, while
S. lanceolata tends to grow in open places such as roadsides, fence-rows,
and power line rights-of-way. Although this sounds like a long list of
distinctions, some botanists would disagree with it entirely, insisting
that these characteristics are not reliable. If you find enough specimens
of each species, you can make your own decision.

S. lanceolata

S. marilandica

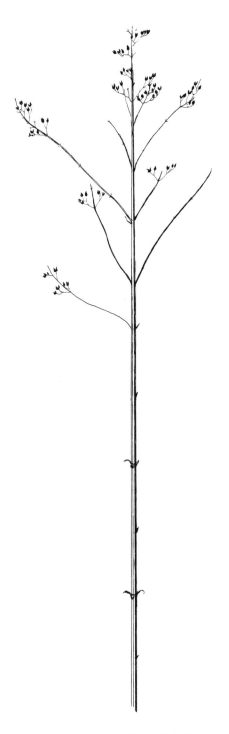

Beard ~ Tongue
Penstemon Digitalis

This perennial grows to a height of four feet in meadows, open woods, old fields, and clearings. In summer, the opposite leaves form almost a cup where they meet. In winter, if no leaves are left, you can still see a little bulge at the nodes where the cup was. The fruit is a two-parted capsule about one-quarter to one-half inch long, and each half has a notch at the tip. The opposite branches stay almost parallel to the main stem, so the plant looks very straight.

DAISY FAMILY
COMPOSITAE

The Daisy family is one of the largest families of flowering plants. In the summertime you can recognize it easily by its daisy-like composite flowers. They look like one flower, but are actually made up of many tiny flowers all together in one head. What you might consider the "petals" of a Daisy—the white parts that you pick off to say, "She loves me, she loves me not"—are actually individual flowers, and the yellow part in the middle is also made up of many individual flowers. These flowers are all attached to a receptacle, *which remains through the winter. The flowers are surrounded by several bracts which also often dry up and remain through the winter.*

The fruit of the plants in the Daisy family is an achene—*a small dry fruit that looks more like a seed. The seed is contained inside this fruit, but the fruit does not open when it is ripe; its outer covering merely rots, leaving the seed to develop. Often, the achene has fine hairs attached to it, collectively referred to as the* pappus. *The "parachute" of a Dandelion fruit is the pappus. (See the glossary for an illustration.)*

Since they are so light, and easily wind-borne by the pappus, the fruits of many of the Compositae do not stick around very long, and we will not use them to identify the winter plants. All that remains of the flower is the receptacle—the surface to which the flowers and fruits were attached—and often the overlapping dried bracts. You can usually spot members of the Daisy family in winter because the dried bracts and receptacle end up looking like a Daisy themselves.

Goldenrod
Solidago spp.

Goldenrods are very common plants that bloom in the late summer and early fall. They often fill entire fields and power line rights-of-way with their bright yellow flowers. Contrary to popular opinion, Goldenrods are not a major cause of hay fever. Hay fever is caused only by plants that release copious amounts of pollen. The only plants that release large amounts of pollen are those that depend upon the wind to spread their pollen to another member of the same species—a rather chancy venture. Brightly colored flowers do not depend on the wind; they depend on insects, who are attracted by their colors. An insect-pollinated flower does not "waste" pollen by spreading it broadcast to the wind; in fact it often hides its pollen at the base of a long tube. Goldenrods are insect-pollinated. In medical tests, their pollen can cause hay fever, but in the outdoor world, it is not responsible, simply because there is not very much of it floating in the air. The real culprit is Ragweed (p. 182), which has small, inconspicuous flowers and is pollinated by the wind. It blooms at the same time as Goldenrod, but nobody notices it.

There are many species of Goldenrod, hard enough to tell apart in summer, let alone winter. In some cases, in winter it is hard to tell a Goldenrod from an Aster. In general, though, Goldendrods grow three to six feet tall, usually in open places. Their branching is usually plume-like, with the flowers usually in dense clusters. They tend to grow in large patches.

Like other members of the Daisy family, their fruit is an achene, with short fuzzy hairs. After it blows away, a flower-like receptacle is left (see detail). You should not try to use the receptacle in identification, as it is quite similar from one species to the next, and quite similar to that of an Aster.

You can preserve flowering Goldenrods by hanging them upside down in a cool, dry place, if you can find one. Make sure they are well ventilated and they will keep their shape and color for several years.

receptacle
with fruits

receptacle
after fruits
have dispersed

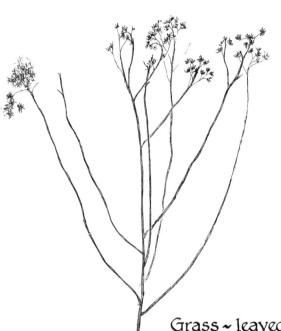

Grass ~ leaved Goldenrod
Solidago graminifolia
(Lance-Leaved Goldenrod)

detail of bracts

This is one of the few Goldenrod species which you can identify easily. It is so different from most Goldenrods, and so common, that it deserves its own identification. You can recognize it by its flat-topped clusters of fuzzy, tan bracts.

You might confuse Grass-leaved Goldenrod with Yarrow (p. 174). However, Grass-leaved Goldenrod is a wispier plant; the stalks are longer, more slender, and more loosely spaced. It is generally larger, growing up to four feet high, while Yarrow rarely exceeds three feet. The bracts of Grass-leaved Goldenrod sometimes have a dark tip, while those of Yarrow are always uniformly tan.

Grass-leaved Goldenrod is a perennial which spreads by underground rhizomes. It grows in open places.

Aster
Aster spp.

In many ways, the Asters are quite similar to the Goldenrods, and it is not always easy to tell one from another. They have a similar fruiting structure and bear similar dried receptacles in the winter (see the next two pages). In general, however, Asters tend to be smaller and more spindly than Goldenrods, and their flowers are more often borne singly. They grow more often in the woods than do Goldenrods, and they rarely fill extensive areas, as Goldenrods do. They are, however, quite common.

As with the Goldenrods, there are many species of Aster which are hard to tell apart even when the plants are in flower. Asters bloom in the fall and brighten up the approach of winter.

The name "Aster" is from the Greek word for "star," referring to the shape of the flowers. This could also apply to the shape of the dried bracts and receptacles.

The height varies from one to seven feet, but most asters grow about two to three feet high.

(see the following pages)

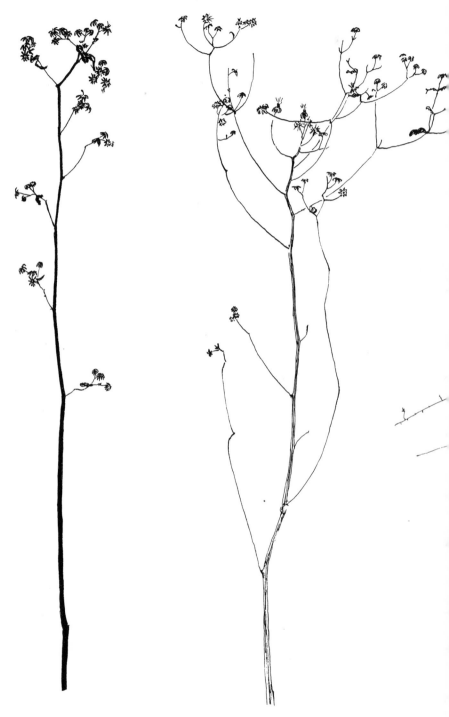

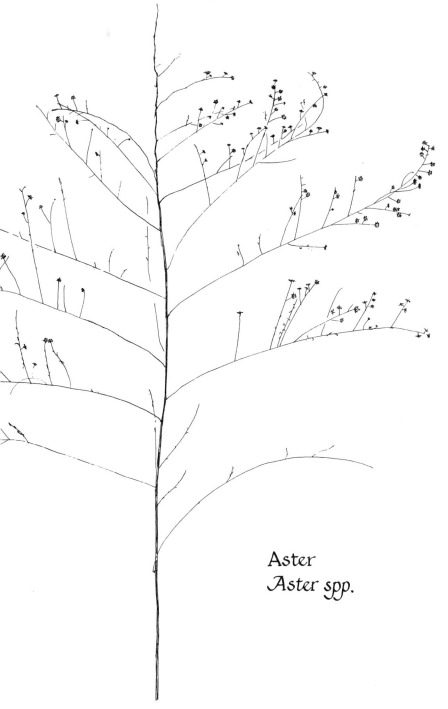

Aster
Aster spp.

Yarrow
Achillea Millefolium

Yarrow is a common perennial of roadsides and fields that grows about one to three feet high. You can recognize it by its flat-topped clusters of tan bracts. Yarrow is often used in architectural models to represent trees. The architects like it because it branches so frequently that they can cut off any number of sections and the plant will still conform to the scale of their model.

Black~eyed Susan
Rudbeckia hirta

Black-eyed Susan is easy to recognize by
its dried cone-like or button-like flower
heads. The stems, bracts, and shrivelled
leaves are all covered with rough hairs.
The plant is very common in meadows
and waste places, and on roadsides. It usually grows two to three feet tall and rarely
has more than a few branches.

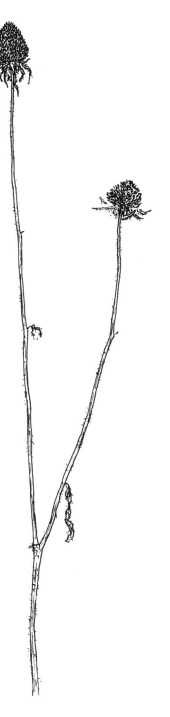

Chicory
Cichorium Intybus

Chicory is another nondescript but very common winter weed. In the summer you have almost certainly noticed its bright blue daisy-like flowers growing along the side of the road. The plant is also common around barns and in pastures, where it can grow almost six feet tall. Branching is variable; you might find plants with many branches or with few. Chicory is a perennial which has a large taproot. In the South and in Europe, the root is roasted, ground up and added to coffee for a taste that some people love and some hate. You can try doing this with roadside chicory, although it is doubtful whether the end product is worth the human and fossil fuel energy you put into it. Endive, *Cichorium endiva,* is another species of chicory, and is sometimes referred to as such.

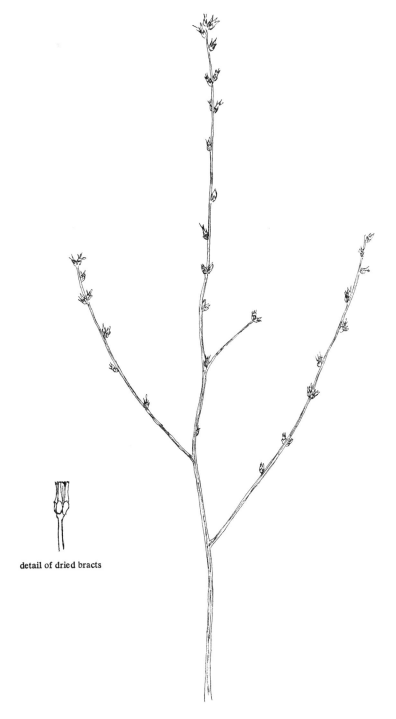

detail of dried bracts

— pappus

— achene

cross-section of bracts
showing fruits inside

Knapweed
Centaurea maculosa

(Star Thistle)

The easiest way to recognize Knapweed is by the distinctive fringed black tips of the lower bracts (see detail). The dried bracts form a cup, inside of which you might find many fine hairs. At the base of the hairs are the fruits (achenes). The overall growth form of Knapweed is variable. It is often much branched and rather scraggly. It grows in the worst soils, and is common in old gravel pits and along roadsides. It is a biennial, growing up to six feet tall.

There are many species of *Centaurea,* of which *C. maculosa* is the most common. The various species are easy to tell apart, because each has a differently patterned fringe on the bracts, but this distinction is beyond the scope of this book. If you have another species of *Centaurea* to identify, look at *Peterson* or *Britton and Brown.*

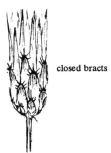

closed bracts

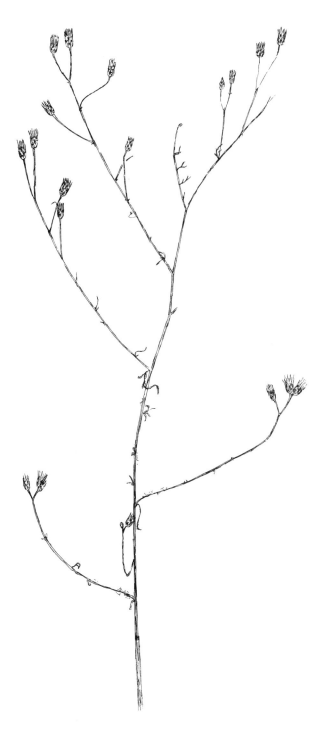

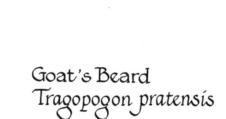

Goat's Beard
Tragopogon pratensis

The best way to describe Goat's Beard is to say that
it looks like a giant Dandelion gone to seed. Goat's
Beard blooms inconspicuously in the early spring, and
the seed is set by May or June. The seeds have all
blown away and the plant has disappeared by fall, so
this is not really a winter weed, but you are bound to
be curious about it if you see it, so it has been included in this book. Goat's Beard is a perennial, growing about one or two feet tall. One of its appealing
characteristics is that you do not have to go very far
to find it; it grows along roadsides and in vacant lots.
If you want to keep a Goat's Beard fruiting head for
decoration, it will hold together better if you spray
it lightly with some kind of fixative.

Tansy
Tanacetum vulgare

You can recognize Tansy by its flat, round
flower heads arranged in a flat-topped
cluster. The plant grows one to three feet
high. If the plant grows near you, you
have probably noticed its bright yellow
button-like flowers in the summer. The
summer plant has finely cut compound
leaves which often dry and remain
through the winter, and these can
help you in identification. Tansy was
originally brought from Europe to
be grown as an herb. The leaves
and the flower heads have a strong,
somewhat smoky smell, and if you
crush the dried flower heads you can
still find some scent. Since its introduction, Tansy has escaped and established itself quite well in fields,
in vacant lots, along railroad tracks,
and in other waste places. Although
its range covers the entire U.S., its distribution is patchy. For instance, it is very
common around the Boston area, but I
have found only isolated plants in Connecticut, the adjacent state.

 dry flower head

Common Ragweed
Ambrosia artemisiifolia

Ragweed does not have much to distinguish itself by, in summer or winter. Its pollen, which is in the air from August to October, is one of the main causes of hay fever, and *Gray's Manual of Botany* calls it "a despised weed." Since there are no really discernible dried flowers or fruiting structures, the best way to identify Ragweed is by its generally raggedy aspect and by its alternate or opposite up-curving branches. If you can find a fruit still on the plant, you will recognize it easily enough by its knobby protuberances (see detail of drawing).

Two species of Ragweed are common in our area. Common Ragweed (*Ambrosia artemisiifolia*) grows up to three and a half feet tall and has a relatively smooth stem. Great Ragweed (*A. trifida*) grows as high as fifteen feet and is generally coarse and bristly in texture. Both species are annuals, growing in vacant lots, along roadsides, and in gardens. They are what is known as "invader species"—those that first inhabit an area that has been stripped, burned, cleared, or somehow disturbed. Our present way of life creates plenty of these areas, so there is little hope for hay fever sufferers.

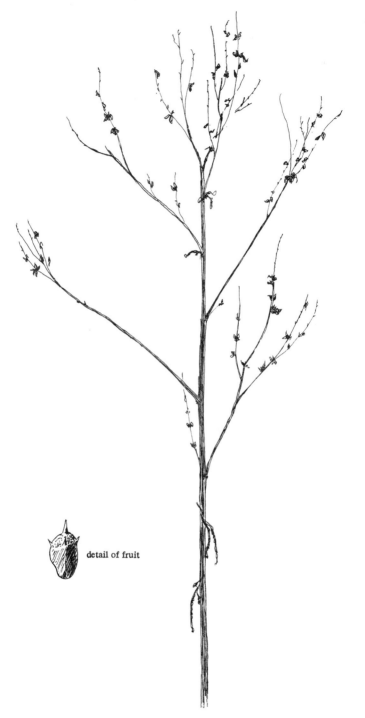

detail of fruit

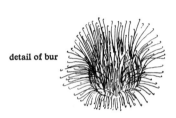

detail of bur

Burdock
Arctium minus
(Clotbur)

You cannot escape noticing Burdock if you ever go near it, because its burs stick tenaciously to your clothes and even to your skin. If you have a cat or a dog, you have doubtless found Burdock burs entangled in its fur. The burs are round and covered with hooked barbs, which used to be the bracts underneath the flower. Burdock can grow frighteningly tall—up to five feet—and its branches spread so widely that the plant's width almost equals its height. It is very common around farm buildings, in dumps, and in other waste places.

Cocklebur
Xanthium strumarium
(Clotbur)

Cocklebur has an oval bur covered with hooked prickles. Each prickle might also be covered with bristly hairs. At the end of the bur are two horns that are not hooked. Cocklebur is a coarse annual that grows from eight inches to six feet tall in all kinds of places—in fields, vacant lots, flood plains, and frequently along lake or sea shores. The stem is sometimes sharply angled, but it is not square. You might confuse it with Burdock (p. 184), but the burs are oval, not round, and they are borne in the axils of the branches, not at the tips. Burdock burs do not have two horns like Cockleburs, and the Cocklebur branches do not spread as widely as those of Burdock.

If you have a plant that almost, but not quite, matches this description, do not worry. For instance, if your burs are oval and borne in the axils, but have no horns at the end, you have another species, *Xanthium spinosum*. If your burs have more or fewer prickles, or are more or less hairy than those illustrated here, the plant is still *Xanthium strumarium*. This is a highly variable species, and botanists argue about whether it should be considered one species or fourteen. For our purposes, it will suffice to treat all of these variations as part of one species.

If you read the description of the Daisy family on p. 167, you might be puzzled by the structure of this plant. The spines are the former bracts, which completely enclose the flower. In winter, if you split open the bur (you will need a sharp knife), you will find it divided into two sections. In each section is a thick black achene.

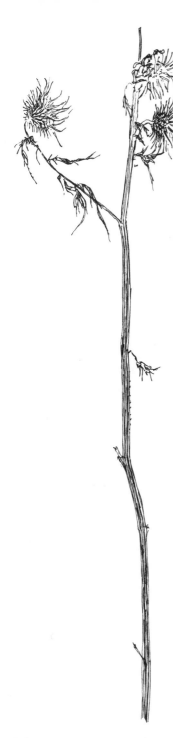

Thistle
Cirsium spp.

Thistles grow to a height of four or five feet.
They are common in fields, on roadsides, and
in pastures, where the cows gladly leave them
alone. The flower heads are relatively large
and made up of many pointed, sometimes
even prickly, bracts. Many of the common
Thistles are biennials, and if you walk bare-
foot through a field in farm country, you will
probably find to your sorrow the rosettes of
the first-year plants. Thistle leaves are prickly,
and in winter you can often find their remains still on the stem.

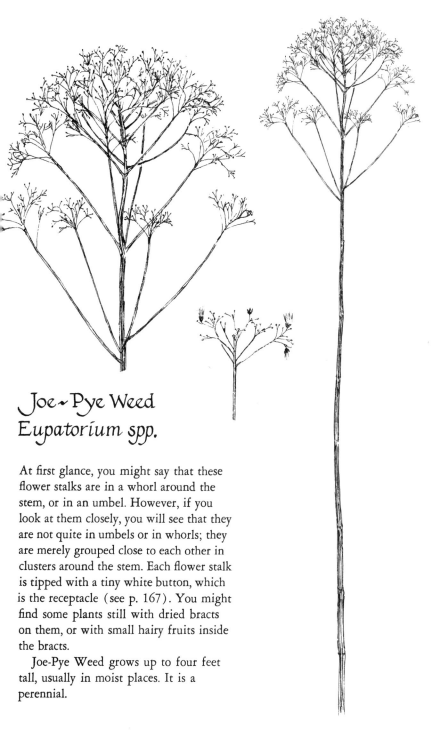

Joe~Pye Weed
Eupatorium spp.

At first glance, you might say that these flower stalks are in a whorl around the stem, or in an umbel. However, if you look at them closely, you will see that they are not quite in umbels or in whorls; they are merely grouped close to each other in clusters around the stem. Each flower stalk is tipped with a tiny white button, which is the receptacle (see p. 167). You might find some plants still with dried bracts on them, or with small hairy fruits inside the bracts.

Joe-Pye Weed grows up to four feet tall, usually in moist places. It is a perennial.

Sweet Everlasting
Gnaphalium obtusifolium

(Catfoot)

If you crush these flower heads, they smell like tobacco. The stem of this plant is very woolly. The dried leaves, which usually stay on the plant all winter, are slightly so. The leaves are long and thin, often brown on the underside. Perhaps because of the wind-blown look of the leaves, and because the wool of the stem gets matted down by the rain, the plant is usually quite bedraggled in winter. A perennial, Sweet Everlasting grows one to two feet tall in dry sandy soil. You can distinguish it from Pearly Everlasting (p. 192) by its more brownish color and its smell.

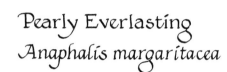

Pearly Everlasting
Anaphalis margaritacea

This species grows mainly in northern areas and in the mountains. You can recognize it by its round flower heads made up of pure white papery bracts, and by its woolly leaves and stem. The plant looks almost the same whether it is dead or alive, for the white bracts are its most prominent feature at any time.

You might confuse Pearly Everlasting with Sweet Everlasting (on p. 190). However, Pearly Everlasting is a really pure white, while Sweet Everlasting is somewhat dingy. The crushed flower heads of Sweet Everlasting have a strong smell, while those of Pearly have no smell at all. To a limited degree, you can also tell the two apart by their ranges and habitats .Although both grow in dry soil, Pearly Everlasting is more common in northern areas and in the mountains—in fields and along roadsides. Sweet Everlasting is found more often to the south and grows in extremely dry, sandy soil.

Pearly Everlasting is a perennial that grows one to three feet high.

Pussy~Toes
Antennaria spp.

(Ladies' Tobacco, Everlasting)

Pussy-toes produces a few drooping flowers at the end of a scaly stalk. The stalk rises up from a rosette of fuzzy green leaves, which you can usually find still alive in the winter. The plant spreads as the leaves creep along the ground and put down new roots. Pussy-toes is quite common in dry, rocky places, but it is inconspicuous because of its small size and dull colors. It rarely grows more than a foot high.

Pussy-toes is adapted in many ways to living in dry places. The hairs on the leaves cut down on moisture loss, and because the leaves are low on the ground there is only one surface from which water can evaporate. Also, the roots form mats underground which can trap the little water available in the soil.

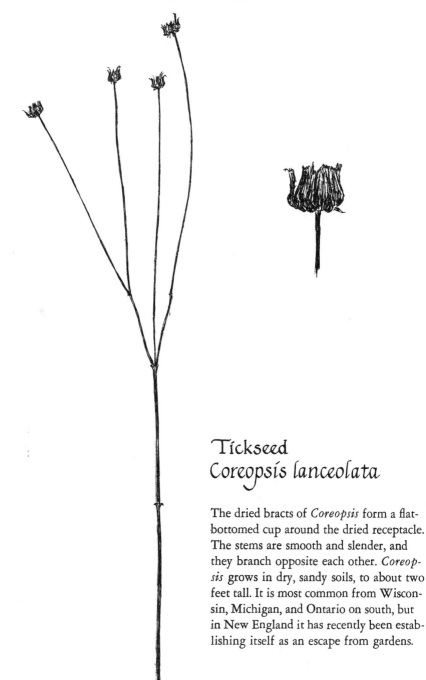

Tickseed
Coreopsis lanceolata

The dried bracts of *Coreopsis* form a flat-bottomed cup around the dried receptacle. The stems are smooth and slender, and they branch opposite each other. *Coreopsis* grows in dry, sandy soils, to about two feet tall. It is most common from Wisconsin, Michigan, and Ontario on south, but in New England it has recently been establishing itself as an escape from gardens.

Sneezeweed
Helenium autumnale

You can recognize this plant by its glob-
ular heads of dried bracts and its rough,
winged stems. It grows in thickets,
swamps, and wet meadows, to a height of
two to five feet.

Elecampane
Inula Helenium

This is an elegant-looking perennial that often grows as tall as five feet. You can find it along roadsides, in rich clearings, and in pastures. Its receptacle is flat, and large compared to that of most composites— about one-half inch to one inch across. You can plainly see the pockmarks left by the composite flowers. The receptacle is surrounded by long, thin, shiny bracts, and these are subtended by more rounded, overlapping bracts. The stem is often sharply angled, but it is not square.

The English name is a corruption of the Latin name. In medieval times, the plant was called *Enula campana,* "Inula of the fields." This gradually became "Elecampane." *Inula* is much simpler.

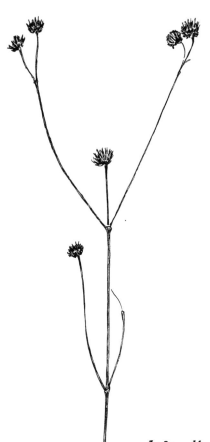

Woodland Sunflower
Helianthus divaricatus

This is a plant of dry soil and unfavorable
environments. You often find it growing
along ridge-tops or in dry shoreline woods.
It has opposite branching, and its stem has a
rough texture. At the tip of each branch is a
head of dried bracts. It grows to a height of
two to six and a half feet.

MISCELLANEOUS PLANTS

Cat ~ tail
Typha latifolia
T. angustifolia

detail of fr

Cat-tails are easy to recognize by their densely packed, cigar-shaped spikes. The spikes are made up of countless miniscule rust-colored fruits, each one of which is attached to the stem by an equally minute stalk. The stalk has many fine hairs on it which are packed down while the fruit is maturing. When the fruit is ripe the spike "explodes" and the hairs fluff out and carry the tiny fruit away.

If you want to bring Cat-tails inside for decoration, pick them as early in the season as you can; late in the summer is best. You should dip them in lacquer or spray them with some kind of fixative; otherwise they will explode, and you will have quite a mess to clean up.

Cat-tails grow up to nine feet tall. They spread by underground rhizomes and often cover large areas of fresh or brackish water marshes. The shores of the Quinnipiac River, north of New Haven, Connecticut, are lined with a thousand acres of solid Cat-tails. Cat-tail marshes make an excellent habitat for muskrats, and you will often see the animals' round houses made of Cat-tail stems and leaves. The muskrats also eat the Cat-tails.

There are two common species of Cat-tail in the Northeast: *Typha latifolia,* which has broad leaves, and *T. angustifolia,* with more narrow leaves. In summer, their flowering structure makes it easy to tell them apart, but in winter, you must rely on the relative size of the leaves, or of the fruiting spikes, to tell the two apart. This is not always easy, if you don't have one of each for comparison, but it is not worth worrying about.

Bur ~ reed
Sparganium spp.

Bur-reed grows in mud and shallow water. Although the plant stands erect when it is alive, in winter you will probably find the dead stalk floating limply in the water. The small, pointed fruits are all packed into a round, bur-like head an inch or less in diameter. The leaves look like a ribbon; in fact, the Latin name comes from the Greek word for "swaddling band."

empty sepals

fruit

Water Plantain
Alisma spp.

clusters of fruit

This plant is distinguished by having a triangular stem. Its flower stalks are in whorls of three to ten stalks which divide once or twice again into further whorls. The small, flat fruits are rounded on the back and arranged in circular clusters inside the three pointed sepals. The fruits fall off during the winter, leaving the sepals. *Alisma* grows up to three and a half feet high in wet places, usually where there is standing water.

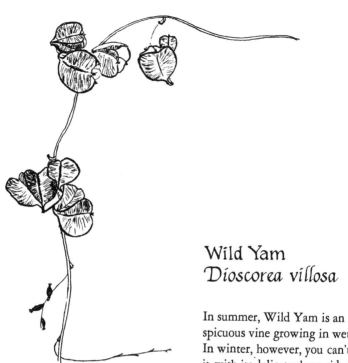

Wild Yam
Dioscorea villosa

In summer, Wild Yam is an inconspicuous vine growing in wet places.
In winter, however, you can't miss
it, with its delicate three-sided, heart-
shaped fruits. The fruits are tan,
about one-half inch to one inch
long, and might still be filled with
light, flat seeds.

Wild Yam is in the same genus
as the cultivated yam, *Dioscorea
alata*. A yam is not the same thing
as a sweet potato (*Ipomoea Batatas*),
although the two are often confused.
Sweet potatoes are more commonly
grown and marketed in the U.S., but
there are two types, one of which is
sweeter than the other and is often
called a yam. Sweet potatoes and
yams are both staples in tropical
countries.

Pigweed
Chenopodium spp.

(Lamb's Quarters, Goosefoot)

This plant does not have much to distinguish itself by. It is scraggly, usually three to four feet high, and quite variable in its branching habits. It is an annual weed that grows in gardens, in vacant lots, and along roadsides. If you find any seeds on your plant, you might notice a similarity between them and the seeds of spinach, beets, or chard: all of these plants are in the same family. If you plant your spinach late, and it goes to flower before you get any leaves to eat, you will see that the whole plant looks somewhat like Pigweed. One species of Pigweed, *Chenopodium album,* is delicious to eat in the early spring. There are many species of *Chenopodium* in our area, but they are hard enough to distinguish in summer, let alone in winter.

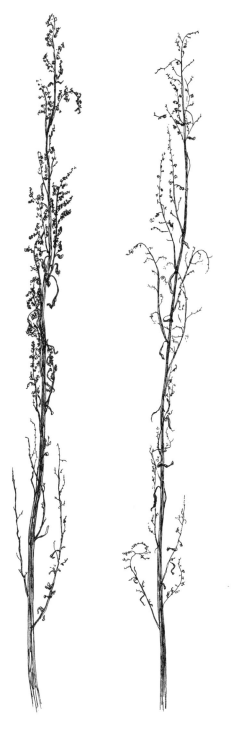

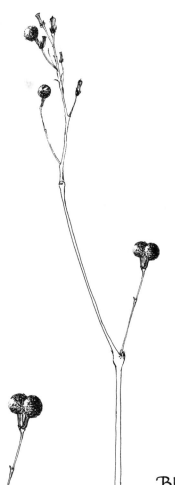

Blue Cohosh
Caulophyllum thalictroides

Blue Cohosh is easy to recognize by its deep
blue berries. If they are all missing you can
still identify the plant by its shiny tan stem,
which bulges at the nodes. The hourglass-
shaped flower stalks remain on the plant and are
also helpful. Blue Cohosh grows up to three and
a half feet tall in rich, moist woods. South of
New England it confines itself mainly to the
mountains.

Sundew
Drosera spp.

In the field, you can recognize Sundew most easily by its rosette of distinctive basal leaves. They are covered with gland-tipped hairs that look like dewdrops, whence the name of the plant. These glands are sticky, and if an insect lands on the leaf it becomes trapped. The glands then secrete chemicals which digest the insect, and thus the Sundew plant obtains its nourishment. Sundew is the only carnivorous plant in this book.

If you have found some Sundew stalks without leaves, you can still recognize the plant by its long, three-parted capsules arranged in a short spike. The flowering stalk is usually only a few inches high.

Sundew is a plant of bogs and swamps. It is predictably found in acid bogs such as those of the New Jersey Pine Barrens—the kind of bog where cranberries grow. However, a little low spot in an upland meadow might also support a stand of them.

There are several species of Sundew, distinguished mainly on the basis of leaf shape. Many have more narrow leaves than those illustrated here, but they all have the same distinctive glands.

Ditch-Stonecrop
Penthorum sedoides

This plant is fairly inconspicuous in summer, but it is more likely to
attract your attention in winter with its tiny, amazingly complex,
reddish-brown capsules. Each capsule is divided into five sections on top
of a flat, star-like dried calyx. When closed, each one has five horns,
and the fruit opens by the falling off of these horns. When opened,
the inside of the fruit is an off-white shade. Less than a quarter of an inch
in diameter, the fruits are all borne on the upper side of the branches
that arch out from the top of the stem.

Ditch-Stonecrop is a perennial which spreads by rhizomes. It grows
about one to three feet tall, in open, wet places. The Latin name
Penthorum is from the Greek *pente,* "five," referring to the five sections.

closed fruit

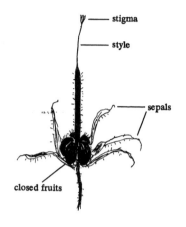

stigma

style

sepals

closed fruits

Cranesbill
Geranium maculatum

Although this curious fruit does not last into the winter, it is so distinctive that it deserves to be included in this book. The long, vertical beak is the style of the flower, the part that leads from the stigma to the ovaries. The five bristles at the tip of the style are the dried stigmas, the sticky part of the flower on which the pollen lands. At the base of the long style are the fruits. Each fruit has a long beak which is attached to the whole length of the style, but when the fruits are ripe they break off and curl outward from the style, staying attached only at the tip.

You should look for these fruits in June or July, in woods and meadows. The plant grows about one to two feet tall and is quite common in most areas. Since the fruit is usually set by late June, you will still find leaves on the plant which will help you to identify it as a *Geranium*. All parts of the plant are hairy. *Geranium maculatum* is the most common species of wild *Geranium*, but there are many others, with similar fruits. The English and the Latin names refer to the shape of the style, which apparently looks like the beak of a crane. The household geranium is a different genus, in the same family—*Pelargonium*.

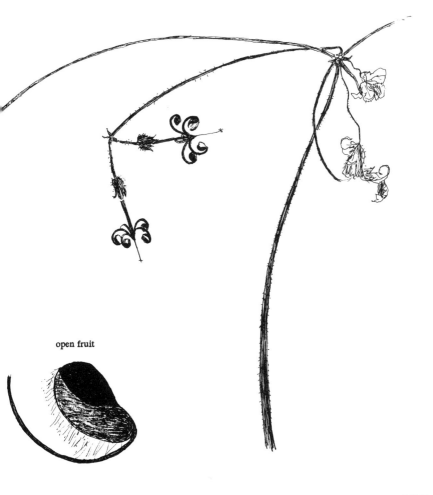

open fruit

Bittersweet
Celastrus orbiculatus

The Bittersweet fruit has a yellow or orange outer layer which opens into three sections to expose the bright red-coated seeds. The fruit is about a half-inch in diameter. Bittersweet is a popular dried plant that you often see for sale in gift shops and farm stands. It is silly to spend money on it when you can find it for yourself in the field. If you want to collect it, there is no special kind of place to go to, for Bittersweet has a variety of habitats—roadsides, thickets, riverbanks, and woods. It seems to prefer edges; you will rarely find it deep in the forest. When you go, be sure to take a knife or some shears with you, for the stems are hard to break, and be sure to collect it early in the fall. After Thanksgiving, most of the outer coverings will have fallen off, and the berries will be shrivelled and unattractive. If you are so early that the outer covering is still closed, you can collect the plants anyway, and the fruits will open up when you bring them inside.

There is some indication that Bittersweet is disappearing from the countryside due to the eagerness of commercial collectors. For this reason, if you want to pick it, you should try not to take too much. If you want an endless supply, you might try planting it in your garden. You can buy young plants at a nursery.

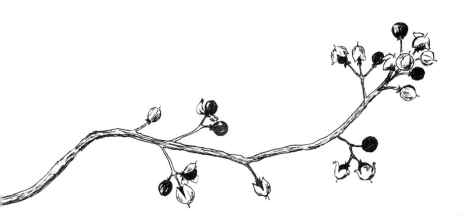

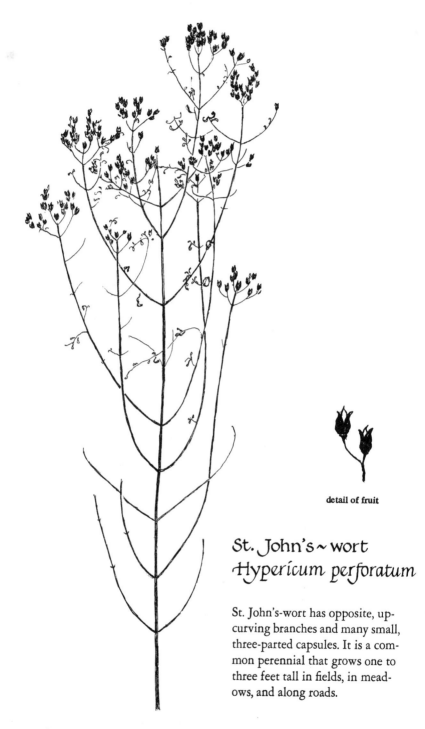

detail of fruit

St. John's~wort
Hypericum perforatum

St. John's-wort has opposite, up-curving branches and many small, three-parted capsules. It is a common perennial that grows one to three feet tall in fields, in meadows, and along roads.

detail of fruit

Pineweed
H. gentianoides

Pineweed is a tiny annual plant usually no more than six inches tall. It has many wiry, opposite branches. The fruits are hard to see because they are so small, but if you look closely you will see that they are three-parted capsules like those of its relative, the common St. John's-wort, on the preceding page. *Hypericum gentianoides* grows in places with dry, sandy soil, such as railroad embankments and old gravel pits.

closed fruit

remaining bracts

Pinweed
Lechea spp.

Perhaps the English name of this perennial plant comes from the fact that from a distance the fruits look like pinheads. Up close, you will see that they are three-parted capsules, some of them opened, some still closed.

The plants grow four inches to one and a half feet high, with many fruits. Pinweed grows in dry, sandy soil, but it is not an urban vacant lot plant. You find it more often along abandoned railroad rights-of-way and in sandy woods. In summer the plant is exceedingly inconspicuous, for the flowers are tiny. You could walk past it many times without noticing it and then see it clear as day in the winter, when all the other greenery around it is gone.

There are several other species of *Lechea* that closely resemble this one. Some have slightly larger fruits and some have a slightly different growth form. They are not very common, and they are hard to tell apart from each other. So if you have a plant that almost, but not quite, matches the above description of Pinweed, you have probably found one of these other species.

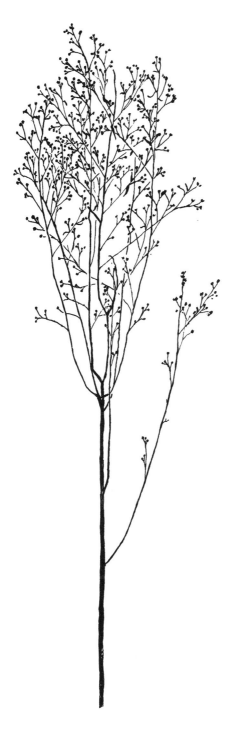

Purple Loosestrife
Lythrum Salicaria

Purple Loosestrife grows very tall, four to six feet and branches like a candelabra. Its fruit is a small capsule more or less arranged in whorls. The fruits disintegrate very easily, however, and you will often find nothing left but stubs and stalks. It is easier to recognize the plant by its height, its general form, and its habitat. Purple Loosestrife was introduced from Europe and has been quite successful in establishing itself. It grows in wet, open marshes, and in summertime often covers huge expanses with its lurid purple-red flowers. If you ever take the shoreline route, either by car or train, from New York to Boston in late July, you can see mile upon mile of purple marshes all through Rhode Island and eastern Connecticut.

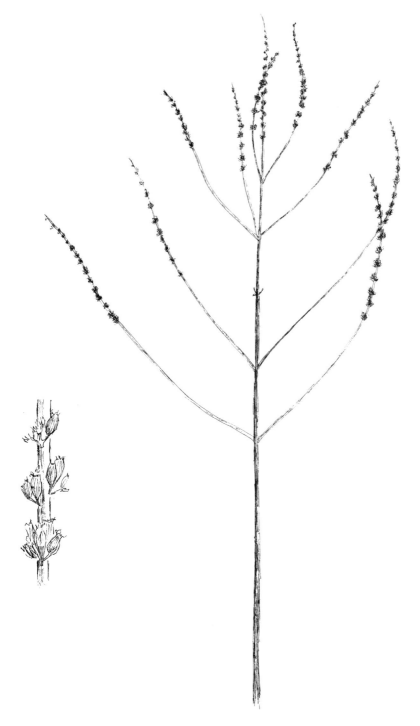

Loosestrife Family / Lythraceae **221**

Water Willow
Decodon verticillatus

The calyx, which covers almost the entire fruit,
has many irregular, delicate, curved teeth which
make the whole fruit look frilly. The capsules
are usually divided into three main parts (sometimes four or five), and they are round, but
somewhat squashed at the bottom. They are half
an inch or less in diameter. The branching pattern of the flower stalks is distinctive.

Water Willow grows along the shores of
lakes and streams, and you can recognize it from
a distance by its arching stems. Young shoots
arise out of the water and arch back in; where
they touch the bottom, they send out new roots.
From these new roots, a new shoot arises and
also spreads; thus the plant forms large patches.
The stems grow anywhere from one to three
yards long. Like all plants that spread vegetatively, Water Willow is a perennial.

Evening Primrose
Oenothera biennis

These fruits are long, four-parted, woody capsules
arranged in a spike at the tip of the stem. The
plant grows three to six feet tall and is very common in fields, fill areas, old gravel pits, and other
waste places. As the Latin name implies, it is a
biennial. Near this year's dead stalks, you can easily
recognize the rosettes of leaves that will send up
next year's flowering shoot. The leaves are long
and toothed, each one tinged with a spot of red.

detail of fruit

Seedbox
Ludwigia alternifolia

Seedbox is one of the easiest plants in this
book to recognize, for it is the only plant
with a square fruit. The fruit opens by a pore
at the top, and the seeds are shaken out by the
wind throughout the winter. It is an amazing
piece of work. You can find Seedbox in wet
places, particularly fields. It grows one to
three feet tall as a perennial.

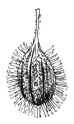

Enchanter's Nightshade
Circaea quadrisulcata

Although this plant can grow as tall as three feet, its fruits are tiny—never more than a quarter of an inch long. The fruits are a distinctive teardrop shape, with three to five grooves on their surface. (The specific Latin name, *quadrisulcata,* means "four-furrowed.") The fruits are covered with stiff little bristles and tend to droop downward from the stem.

Enchanter's Nightshade grows as a perennial in rich woods, thickets, and ravines. You almost always find it growing in patches, so once you have found one plant, look around and you will find many more.

open fr

 closed f

Whorled Loosestrife
Lysimachia quadrifolia

Yellow Loosestrife
Lysimachia terrestris

 dried se

(Swamp Candles)

These two species have similar fruits—small, round capsules that open into three to six sections. The capsules often fall off and leave five or six pointed sepals. The two species also grow about the same height—one to three and a half feet, but they are still easy to tell apart. The fruiting stalks of Whorled Loosestrife are found in the leaf axils, while those of Yellow Loosestrife are all above the leaves. (Even if no leaves are left, you can still see their scars.) Also, the stalks of Whorled Loosestrife, as well as the leaves, are borne in true whorls. Those of Yellow Loosestrife are almost, but not quite, in whorls, and the leaves are in pairs rather than whorls. Whorled Loosestrife can be found in many habitats—woods, thickets, roadsides, and power line rights-of-way—but Yellow Loosestrife grows only in swamps.

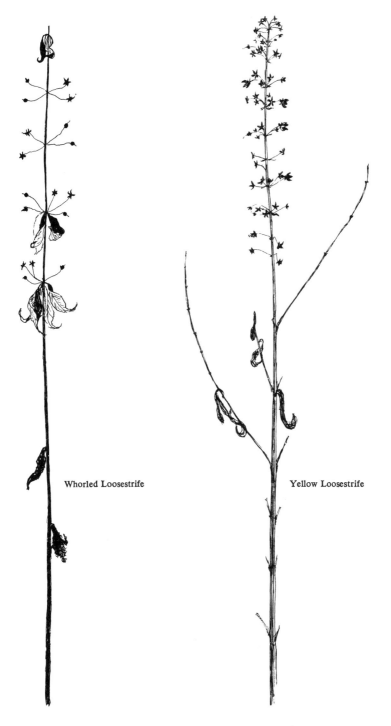

Whorled Loosestrife

Yellow Loosestrife

Sea Lavender
Límoníum spp.
(Marsh Rosemary)

Sea Lavender is one of the most beautiful plants in this book. It grows
in salt marshes up and down the Atlantic coast, and its lavender sprays
can tinge a whole marsh in late summer. If you pick it while it is still
blooming, the flowers will keep their lavender color for a few weeks,
even if you put them in a vase without water. Gradually the lavender
flowers will die, but the stems and the five-toothed calyx will remain
for years. You can recognize the plant by its bushy aspect and by the
dried flowers all lined up on one side of the stem. The plant is a perennial, growing eight inches to two feet tall.

The two species, *L. Nashii* and *L. carolinianum,* are quite similar, and
there is a possibility that the differences between them are a result of
geography rather than genetics. *L. carolinianum* does not grow north of
Long Island, while *L. Nashii* can be found from Florida to Newfoundland.

Dogbane
Apocynum androsaemifolium

(Indian Hemp)

Although Dogbane is not in the same family as Milkweed (p. 131), it is closely related, as you can see from the long woody pods and the small brown seeds with the silky hairs. The thin pods are three to eight inches long and are arranged in pairs. The seeds are very similar to those of Milkweed, except that they are thinner. The Dogbane plant grows one to four feet high and can be found in fields, thickets, or woods.

bracts

Viper's Bugloss
Echium vulgare

(Blueweed)

From a distance, this plant looks
like an indistinguishable fuzzy
mass. The best way to recognize
it is by its curved, snake-like
branches. In the botanical literature, these are referred to as
"scorpioid cymes." The plant
grows along roadsides and in
waste places, and reaches a height
of three feet.

If you look at the plant closely,
you will see many long, pointed
bracts lined up along one side of
each branch. The hairs that cover
the whole plant are bristly, almost
prickly. In summer, the flowers
are a pretty blue, and you might
sometimes find a dried blue
corolla still hanging from the
plant.

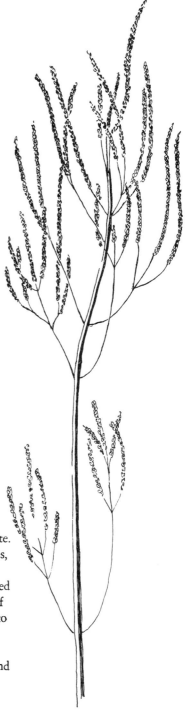

Blue Vervain
Verbena hastata

The stem of this plant is square, and the
branching is almost, but not quite, opposite.
The fruits, which are arranged in tall spikes,
are tiny nutlets which you can hardly see.
They are enclosed in the dried four-toothed
calyx, but this is also hard to see because of
its small size. The whole plant grows one to
five feet high, in swamps, moist fields, and
meadows.

Vervain is a perennial, and you usually find
it growing singly, rarely in patches.

Beech-drops
Epifagus virginiana

These little plants are parasitic on the roots of Beech trees, so you will never find them growing anywhere except under Beech trees. They grow up to one and a half feet tall. The dead stalks are brown and the branches somewhat arched. If you go to pick one of the plants you will probably inadvertently pull up the roots along with it, and they are very distinctive—short and woody, emanating from little knobs.

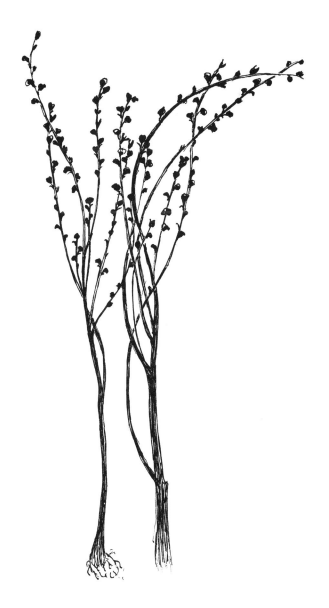

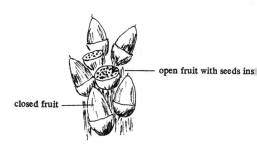

open fruit with seeds ins[

closed fruit

Common Plantain
Plantago major

English Plantain
Plantago lanceolata

Plantains are very common weeds in lawns and gardens, and in winter you can still find rosettes of simple basal leaves. The fruiting stalks can grow as high as two feet, but you rarely find them growing more than six to eight inches high. The plant has many small fruits closely packed in a spike. The fruit is hard to distinguish, but if you look closely, you will see that it is a conical capsule which opens across its middle. When the top half falls off you are left with the rounded bottom half, which contains the seeds.

In the Northeast we have two common species of Plantain. Although their fruits are slightly different, it is easy to tell them apart simply by their general form. The fruits of *Plantago lanceolata* are all crowded into a small spike at the tip of the stem, while those of *P. major* run almost the whole length of the stem.

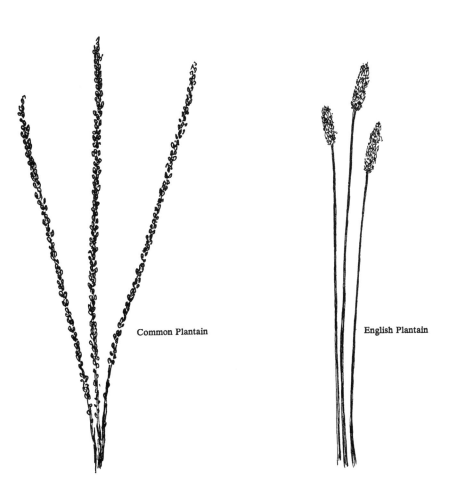

Common Plantain

English Plantain

Plantain Family / Plantaginaceae **237**

Teasel
Dipsacus sylvestris

You can recognize Teasel easily by its spiny, egg-shaped flower heads. The bracts that curve up underneath the head are also spiny, as are the stems. Although Teasel is rare in New England, it is a common roadside plant in parts of New York, Pennsylvania, and New Jersey, where it grows as high as six feet.

Teasel has been used in the production of cloth, although there is some dispute as to exactly how. Some authors say that it was used for fulling, which is the process of beating cloth to thicken it. A different species, *Dipsacus fullonum,* or Fuller's Teasel, was apparently specially grown for this purpose, and you can still find *D. fullonum* growing near old wool-mills. However, a Teasel head seems much too fragile to be used for beating something. Other authorities claim that Teasel was used for carding. This is much more plausible, as carding is the process of combing out and untangling the wool to prepare it for spinning. In fact, the name *teasel* and the word *tease* both came from the Anglo-Saxon word *taesan,* which means "to pull to pieces." Although carding machines have been invented, some weavers claim that nothing does the job as well as Teasel. As of the 1950's Teasel was still being raised commercially.

You can grow Teasel in your garden from wild seeds, even if you live where it is not common. If you cannot get any wild seeds, you can order seeds from an herbal or exotic plant nursery. Plant them in a sunny spot. Since Teasel is a biennial, you will have to wait two years before you have a fruit. In the first year it will only produce a rather unattractive rosette of wrinkled leaves with bristly hairs. When you harvest the Teasel heads, be sure to shake the seeds out or leave a few plants standing, and it will probably grow back by itself.

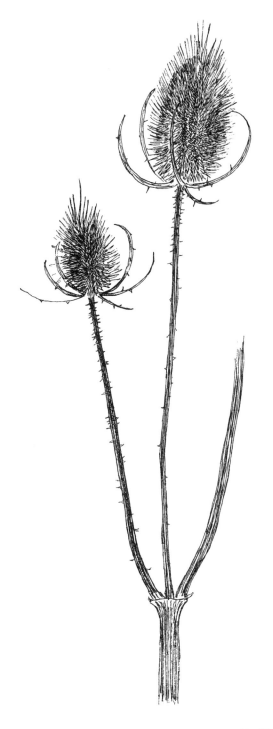

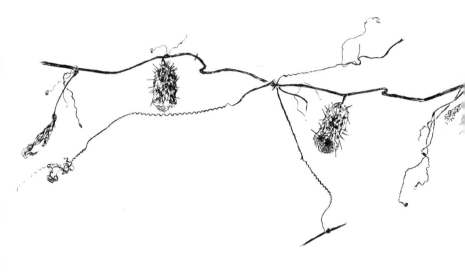

Wild Cucumber
Echinocystis lobata
(Balsam-Apple)

You can't mistake this bizarre vine with its spiny fruits and its frenetic, tightly wound tendrils. The fruit, about one inch long, consists of two layers. The inner layer is netlike and spongy and divided into two or three sections. The outer layer is papery, wrinkled, and covered with weak spines. Look on the ground below where you find a Wild Cucumber vine and you might find some of the inner sponges that have dropped off the plant. A close relative of the Wild Cucumber called *Luffa* is cultivated for its sponges, which people use as scouring pads. Wild Cucumber is also related to the garden cucumber, squashes, pumpkins, melons, and gourds. It is an annual that grows in moist places and thickets.

The Latin name comes from two Greek words: *echinos,* "hedgehog," a reference to the thorns, and *cystis,* "bladder," referring to the shape of the fruits.

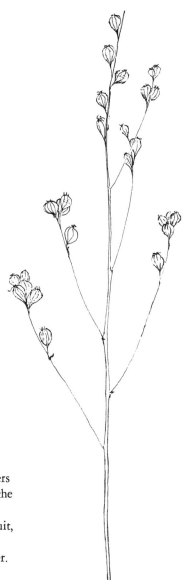

Indian Tobacco
Lobelia inflata

In summer, the Indian Tobacco flowers
are subtended by a tubular calyx. As the
fruit matures, the calyx becomes inflated and completely encloses the fruit,
which is a two-parted capsule. The
papery calyx is what you see in winter.
The calyx is usually less than a half
inch long, and the whole plant grows
eight inches to three feet tall. Indian
Tobacco is a poisonous annual that
grows in many kinds of places—fields,
cleared areas, roadsides, and open
woods. You usually find it growing as
a single plant, rarely in large patches.

INDEX

THE AUTHOR

Lauren Brown is a graduate of Swarthmore College and of the Yale School of Forestry and Environmental Studies. A native of Guilford, Connecticut, she has taught plant identification at Yale University and now lives near New Haven.